AF595109

Physical Vapor Deposited Biomedical Coatings

Physical Vapor Deposited Biomedical Coatings

Editors

George E. Stan
Bryan W. Stuart

MDPI • Basel • Beijing • Wuhan • Barcelona • Belgrade • Manchester • Tokyo • Cluj • Tianjin

Editors
George E. Stan
National Institute of Materials Physics
Romania

Bryan W. Stuart
The University of Oxford
UK

Editorial Office
MDPI
St. Alban-Anlage 66
4052 Basel, Switzerland

This is a reprint of articles from the Special Issue published online in the open access journal *Coatings* (ISSN 2079-6412) (available at: https://www.mdpi.com/journal/coatings/special_issues/PVD-biomedical).

For citation purposes, cite each article independently as indicated on the article page online and as indicated below:

LastName, A.A.; LastName, B.B.; LastName, C.C. Article Title. *Journal Name* **Year**, *Volume Number*, Page Range.

ISBN 978-3-0365-2414-6 (Hbk)
ISBN 978-3-0365-2415-3 (PDF)

Cover image courtesy of Dr. George E. Stan

Contents

About the Editors

George E. Stan received a B.E. degree in Medical Engineering, an M.Sc. degree in Biomaterials, and a Ph.D. degree in Materials Engineering from the University Politehnica of Bucharest, Bucharest, Romania, in 2005, 2007, and 2011, respectively. He is currently a Senior Researcher Rank I at the National Institute of Materials Physics, Magurele, Romania. His research interests are focused on the fabrication of quality thin films via physical vapour deposition techniques on large-area or complex geometry substrates, with applicative range in the field of biomedicine and electronics.

Bryan W. Stuart received a BEng in Mechanical Engineering and a Ph.D in Biomaterials Manufacturing (specifically orthopaedic glass thin films) from the University of Nottingham (UoN), UK, between 2009 and 2017. He continued developing biomaterials via vapour deposition as a postdoctoral researcher before taking up a new position at the University of Oxford, developing industrially scalable flexible electronics via Roll-to-Roll deposition and novel in-vacuum printing techniques. Dr. Stuart left academia for Dyson Technologies Ltd. in 2019, where he worked as a coating engineer for 2 years in energy storage applications. As of 2021, he now works with Oxford Photovoltaics in developing the next generation of solar technology.

MDPI

Editorial

Physical Vapour Deposited Biomedical Coatings

Bryan W. Stuart [1,*] and George E. Stan [2,*]

1 Department of Materials, University of Oxford, Parks Road, Oxford OX1 3PH, UK
2 National Institute of Materials Physics, RO-077125 Magurele, Romania
* Correspondence: bryan.stuart19@gmail.com (B.W.S.); george_stan@infim.ro (G.E.S.); Tel.: +40-21-241-8128 (G.E.S.)

Abstract: This Special Issue was devoted to developments made in Physical Vapour Deposited (PVD) biomedical coatings for various healthcare applications. The scrutinized PVD methods were Radio-Frequency Magnetron Sputtering (RF-MS), Cathodic Arc Evaporation, Pulsed Electron Deposition and its variants, Pulsed Laser Deposition, and Matrix Assisted Pulsed Laser Evaporation (MAPLE), due to their great promise especially in the dentistry and orthopaedics. These methods have yet to gain traction for industrialization and large-scale application in biomedicine. A new generation of implant coatings can be made available by the (1) incorporation of organic moieties (e.g., proteins, peptides, enzymes) into thin films by innovative methods such as combinatorial MAPLE, (2) direct coupling of therapeutic agents with bioactive glasses or ceramics within substituted or composite layers via RF-MS, or (3) by innovation in high energy deposition methods such as arc evaporation or pulsed electron beam methods.

Keywords: physical vapour deposition; thin-films; medical devices; bioactivity; biomimicry

Citation: Stuart, B.W.; Stan, G.E. Physical Vapour Deposited Biomedical Coatings. *Coatings* **2021**, *11*, 619. https://doi.org/10.3390/coatings11060619

Received: 7 May 2021
Accepted: 18 May 2021
Published: 21 May 2021

As research and development into medical devices reaches an all-time high, surface functionalization through less invasive, low dimensional thin-film coatings is at the forefront of optimising bio-integration, bio-activation, and biomechanics. In the last decades, physical vapour deposition (PVD) techniques have gained attraction for their diverse capabilities in blending and manufacturing highly adhesive, novel materials with minimally invasive thicknesses from ångströms to the micrometres scale.

It has long been the opinion of the industrial community that technologies based on thermal spray technologies (e.g., plasma, flame, detonation, cold, high-velocity atmospheric, high-velocity oxy-fuel, and high-velocity suspension flame spraying) and solution-based methods (e.g., sol–gel, solution casting) provide high throughput capabilities and low start-up costs suitable for commercial scale manufacturing. However, as the precision, versatility, scale, and accessibility of standardized and scaled up PVD systems grows, the medical device industry will seek advanced biomaterial applications for the next generation of functional layers with long-term reliability and high success rates.

This Special Issue of Coatings features five full-length articles, all pertaining to the field of bio-functional surfaces with dentistry and orthopaedic applications, spanning from (i) silicon highly doped hydroxyapatite [HA, $Ca_{10}(PO_4)(OH)_2$] coatings [1] and alkali-free copper and gallium co-doped silica bioactive glass [2] layers deposited onto titanium (Ti)-based substrates by Radio-Frequency Magnetron Sputtering (RF-MS), for osseointegration and/or antibacterial uses; to (ii) thin films of HA on polyletheretherketone (PEEK) fabricated by RF-MS [3]; (iii) biological derived HA coatings doped with lithium carbonate or lithium phosphate synthesized by Pulsed Laser Deposition on a 3D printed Ti implants [4]; and to (iv) carbonitride coatings of TiCN, TiSiCN realized by cathodic arc deposition method aiming to improve the wear resistance of the CoCr alloys often used for joint endoprosthetic fabrication [5]. The biofunctionality of the implant-type coatings was surveyed by an array of in vitro (corrosion investigations in biomimetic medium [5], cytocompatibility tests [1,2] and antimicrobial efficacy [2]) and in vivo (implantation into rabbit femoral condyles, followed by mechanical removal at 4 and 9 weeks [4]) biological assays.

Furthermore, this Special Issue suitably demonstrates the current landscape for future generations of biological layers containing two extensive review articles on the emerging popularity of more recent/novel manufacturing methods and their relevance to biomaterial applications. These specifically pertain to the Pulsed Electron Deposition (PED) [6] and Combinatorial Laser deposition technologies [7].

PED is a spark discharge ablation method which accelerates a high energy electron beam towards a target. The functionality has changed over time and most recent developments extract the electrons from a generated plasma in a dielectric tube, termed Pulsed Plasma Deposition (PPD). In an additional development step, ionized jet deposition (IJD), utilizes gas jet entrainment and short pulses of up to megawatts. PPD and IJD exhibit high electrical efficiencies of 30% and 88%, respectively, with higher deposition rates leading to lower capital costs. Of great relevance in current orthopaedic coatings, the high voltage potentials (up to 25 kV) enable deposition of the full spectrum from amorphous transparent film to high energy crystalline phases in a single-step, demonstrated to produce layers such as bone-like HA capable of stimulating osseointegration. Liquori et al. [6] review the application of PED, PPD, and IJD for materials applicable in orthopaedic prosthetics such as wear resistant crystalline yttria-stabilized zirconia coatings or bioactive strontium doped-HA, Bioglass® 45S5, nanostructured silver, and HA-magnetite composite layers onto ultra-high-molecular-weight polyethylene, titanium alloys, stainless steel, glass, silicon, and PEEK substrates.

Axente et al. [7] reviews PLD-based techniques (well-known for preservation of stoichiometry and tune-ability of input energies for desired chemistries/reactability with working gases (e.g., for the deposition of TiN coatings)). The state-of-the-art manufacturing development is described, with emphasis on the Matrix Assisted Pulsed Laser Evaporation (MAPLE) method's capability to fabricate thin (heat-sensitive) organic, inorganic, and hybrid organic-inorganic films. MAPLE relies on UV laser irradiation of a cryogenic target and enables the safe transfer of proteins, biopolymers, enzymes, or polysaccharides, as well as more conventional deposits of calcium phosphates, from a target to a substrate. Some applications include deposition of fibronectin, bovine serum albumin-functionalized graphene oxide nanomaterials. Axente et al. [7] focuses on the "Combinatorial" approach, referring to the use of multiple laser systems to produce material blends across a substrate, demonstrating formation of strontium doped HA–zinc doped β-tricalcium phosphate in bone regenerative applications, sulphated Halomonas Levan blended quaternized low molecular weight chitosan for antibacterial properties, graphene oxide nanomaterials with hybrid bovine serum albumin, and introduced Dabrafenib/Trichostatin A drugs for targeting melanoma cells.

As presented within, PVD modified implant surfaces benefit from enhanced bioactivity, leading to advantageous cell–surface interactions to assist in regenerating tissue. The Special Issue successfully draws attention towards a wide range of PVD methods and novel applications which hitherto have not gained traction for industrialization yet possess enormous benefit for the world of medical care. The methods shown here—Radio-Frequency Magnetron Sputtering, Cathodic Arc Evaporation, Pulsed Electron Deposition and its variants, Pulsed Laser Deposition, and Matrix Assisted Pulsed Laser Evaporation—demonstrated feasible application and great promise as strong advocates for the future widespread adoption of PVD films in dentistry and orthopaedics. A broader field of biomaterials can now be tackled through incorporation of proteins, peptides, enzymes into implant coatings via novel methods such as MAPLE or by the direct coupling of therapeutic agents with bioactive glasses or ceramics within substituted or composite layers via the sputter deposition variants.

Author Contributions: Conceptualization, B.W.S. and G.E.S.; writing—original draft preparation, B.W.S. and G.E.S.; writing—review and editing, B.W.S. and G.E.S.; funding acquisition, G.E.S. Both authors have read and agreed to the published version of the manuscript.

Funding: This work was funded by Romanian National Authority for Scientific Research and Innovation (CNCS-UEFISCDI) in the framework of projects PN-III-P1-1.1-TE-2016-1501 and PN-III-P1-1.1-TE-2019-0463.

Institutional Review Board Statement: Not applicable.

Informed Consent Statement: Not applicable.

Conflicts of Interest: The authors declare no conflict of interest.

References

1. Coe, S.C.; Wadge, M.D.; Felfel, R.M.; Ahmed, I.; Walker, G.S.; Scotchford, C.A.; Grant, D.M. Production of high silicon-doped hydroxyapatite thin film coatings via magnetron sputtering: Deposition, characterisation, and in vitro biocompatibility. *Coatings* **2020**, *10*, 190. [CrossRef]
2. Stan, G.E.; Tite, T.; Popa, A.C.; Chirica, I.M.; Negrila, C.C.; Besleaga, C.; Zgura, I.; Sergentu, A.C.; Popescu-Pelin, G.; Cristea, D.; et al. The beneficial mechanical and biological outcomes of thin copper-gallium doped silica-rich bio-active glass implant-type coatings. *Coatings* **2020**, *10*, 1119. [CrossRef]
3. Hussain, S.; Rutledge, L.; Acheson, J.G.; Boyd, A.R.; Meenan, B.J. The surface characterisation of polyetheretherketone (Peek) modified via the direct sputter deposition of calcium phosphate thin films. *Coatings* **2020**, *10*, 1088. [CrossRef]
4. Duta, L.; Neamtu, J.; Melinte, R.P.; Zureigat, O.A.; Popescu-Pelin, G.; Chioibasu, D.; Oktar, F.N.; Popescu, A.C. In vivo assessment of bone enhancement in the case of 3d-printed implants functionalized with lithium-doped biological-derived hydroxyapatite coatings: A preliminary study on rabbits. *Coatings* **2020**, *10*, 992. [CrossRef]
5. Dinu, M.; Pana, I.; Scripca, P.; Sandu, I.G.; Vitelaru, C.; Vladescu, A. Improvement of CoCr alloy characteristics by Ti-based carbonitride coatings used in orthopedic applications. *Coatings* **2020**, *10*, 495. [CrossRef]
6. Liguori, A.; Gualandi, C.; Focarete, M.L.; Biscarini, F.; Bianchi, M. The pulsed electron deposition technique for biomedical applications: A review. *Coatings* **2020**, *10*, 16. [CrossRef]
7. Axente, E.; Sima, L.E.; Sima, F. Biomimetic coatings obtained by combinatorial laser technologies. *Coatings* **2020**, *10*, 463. [CrossRef]

Article

The Beneficial Mechanical and Biological Outcomes of Thin Copper-Gallium Doped Silica-Rich Bio-Active Glass Implant-Type Coatings

George E. Stan [1,*], Teddy Tite [1], Adrian-Claudiu Popa [1], Iuliana Maria Chirica [1], Catalin C. Negrila [1], Cristina Besleaga [1], Irina Zgura [1], Any Cristina Sergentu [1], Gianina Popescu-Pelin [2], Daniel Cristea [3], Lucia E. Ionescu [4], Marius Necsulescu [4], Hugo R. Fernandes [5] and José M. F. Ferreira [5]

1 National Institute for Materials Physics, RO-077125 Magurele, Romania; teddy.tite@infim.ro (T.T.); adrian.popa@gmail.com (A.-C.P.); iuliana.bogdan@infim.ro (I.M.C.); catalin.negrila@infim.ro (C.C.N.); cristina.besleaga@infim.ro (C.B.); irina.zgura@infim.ro (I.Z.); anyc@yahoo.com (A.C.S.)

2 National Institute for Lasers, Plasma and Radiation Physics, RO-077125 Magurele, Romania; gianina.popescu@inflpr.ro

3 Faculty of Materials Science and Engineering, Transilvania University of Brasov, 500068 Brasov, Romania; daniel.cristea@unitbv.ro

4 Army Centre for Medical Research, RO-020012 Bucharest, Romania; ionescu.lucia@gmail.com (L.E.I.); mariusnecsulescu@yahoo.com (M.N.)

5 Department of Materials and Ceramics Engineering, CICECO, University of Aveiro, 3810-193 Aveiro, Portugal; h.r.fernandes@ua.pt (H.R.F.); jmf@ua.pt (J.M.F.F.)

* Correspondence: george_stan@infim.ro; Tel.: +40-21-241-8128

Received: 12 October 2020; Accepted: 18 November 2020; Published: 20 November 2020

Abstract: Silica-based bioactive glasses (SBG) hold great promise as bio-functional coatings of metallic endo-osseous implants, due to their osteoproductive potential, and, in the case of designed formulations, suitable mechanical properties and antibacterial efficacy. In the framework of this study, the FastOs®BG alkali-free SBG system (mol%: SiO_2—38.49, CaO—36.07, P_2O_5—5.61, MgO—19.24, CaF_2—0.59), with CuO (2 mol%) and Ga_2O_3 (3 mol%) antimicrobial agents, partially substituting in the parent system CaO and MgO, respectively, was used as source material for the fabrication of intentionally silica-enriched implant-type thin coatings (~600 nm) onto titanium (Ti) substrates by radio-frequency magnetron sputtering. The physico-chemical and mechanical characteristics, as well as the in vitro preliminary cytocompatibility and antibacterial performance of an alkali-free silica-rich bio-active glass coating designs was further explored. The films were smooth (R_{RMS} < 1 nm) and hydrophilic (water contact angle of ~65°). The SBG coatings deposited from alkali-free copper-gallium co-doped FastOs®BG-derived exhibited improved wear performance, with the coatings eliciting a bonding strength value of ~53 MPa, Lc3 critical load value of ~4.9 N, hardness of ~6.1 GPa and an elastic modulus of ~127 GPa. The Cu and Ga co-doped SBG layers had excellent cytocompatibility, while reducing after 24 h the *Staphylococcus aureus* bacterial development with 4 orders of magnitude with respect to the control situations (i.e., nutritive broth and Ti substrate). Thereby, such SBG constructs could pave the road towards high-performance bio-functional coatings with excellent mechanical properties and enhanced biological features (e.g., by coupling cytocompatibility with antimicrobial properties), which are in great demand nowadays.

Keywords: implant coating; bioactive glass; copper doping; gallium doping; mechanical; cytocompatibility; antibacterial

---

1. Introduction

The last decades witnessed an unprecedented demand for innovative bio-functional materials, capable of not only preventing failure, but also prolonging the life time of orthopaedic/dental implants and bone grafting scaffolds. Silica-based bioactive glasses (SBG) represent a group of reactive biomaterials that possess outstanding features (e.g., bioactivity, osteoproduction, angiogenesis), which are crucial for assuring an excellent interfacial bonding between the host bone tissue and implants [1–7]. The continuous effervescence in the SBG realm, associated with its unique physical-chemical and biological functional properties [2,5,7], is certified by a progressive yearly increase in the number of publications [5].

Most of the reported SBG compositions investigated so far were inspired by the 45S5 silicate glass (mol.%: SiO_2—46.1, Na_2O—24.4, CaO—26.9, P_2O_5—2.6), developed by Hench and his co-workers [8]. Over 30 years since its first implantation in 1985, 45S5 bioactive glass has been implanted into 1.5 million patients worldwide under various forms and trade names (e.g., Perioglas®, NovaBone®, Biogran®, NovaMin®, TeraSphere®) [2,5]. Due to their high indices of bioactivity, 45S5-based formulations have been proposed as coating materials for the bio-functionalisation of various metallic titanium-based implants [7,9], to enhance their bone-bonding ability. However, the disparity of the coefficients of thermal expansion (CTE) of 45S5 (i.e., ~14–17 × 10^{-6} °C^{-1}) and titanium-based (i.e., ~8.5–9.6 × 10^{-6} °C^{-1}) materials cannot be circumvented with ease [10], and represents a hazard for the long-term mechanical performance of the implant coating. Furthermore, because of the burst release of high concentrations of sodium and overall excessive bio-reactivity [5,11], 45S5 bioactive glass could generate highly alkaline pH environments, which, in the absence of preconditioning procedures could lead to less than favourable biological outcomes [5,12]. More recently, alkali-free SBGs such as FastOs®BG (mol.%: SiO_2—38.49, CaO—36.07, P_2O_5—5.61, MgO—19.24, CaF_2—0.59) have been introduced and acknowledged for their notable performances such as reduced CTEs mismatch, chemical durability, high biomineralization capacity and excellent cytocompatibility [5,9,13].

The possibility to provide new biological functionalities to SBG materials by the inclusion of therapeutic ions within their chemical structure has changed the paradigm in this field, opening new avenues in healthcare and tissue regeneration. Various biologically active elements (e.g., boron, cobalt, magnesium, strontium, zinc) commonly present in the human body have been incorporated in SBGs, as medicinal micronutrients to promote bone regeneration [2,3,5,14,15]. These active ions could influence to some extent the structural properties (e.g., morphology, crystallinity, solubility) of these glasses and their biological responses (e.g., degradability, stem cells differentiation and proliferation, osteogenesis, angiogenesis) [5,16–18]. Upon the dissolution of glasses in the physiological environment, these ions are released and induce additional therapeutic effects.

A growing issue in biomedicine is the increased number of implant failures associated to microbial infection. Thus, it is of paramount importance to provide antibacterial properties to the implant by the addition of specific antimicrobial agents. In this context, various studies have reported the benefit of silver and zinc added as antimicrobial agents in SBGs [5,15,17]. More recently, gallium (Ga) and copper (Cu) have emerged as a potent new generation of antibacterial ions that may be useful in treating and preventing localized infections [19–29]. Valappil et al. [19] reported on the positive bactericidal effects of gallium-substituted phosphate-based glasses (1–5 mol.% Ga_2O_3) against *Escherichia coli*, *Pseudomonas aeruginosa*, *Staphylococcus aureus*, methicillin-resistant *S. aureus* and *Clostridium difficile*. The bactericidal activity due to Ga^{3+} ions was found effective for a Ga_2O_3 concentration as low as 1 mol.% [19]. The potential antimicrobial activity of gallium against *E. coli* and *S. aureus* [22], as well as *P. aeruginosa* [23], has been further confirmed in Ga-substituted SBGs. In relation to CuO-substituted SBGs, recent studies have also demonstrated the potency of Cu to eradicate the *E. coli* [24,25,28], *P. aeruginosa* [28], *S. aureus* [21,25,27], *Salmonella enterica* [28] and *Staphylococcus epidermidis* [25] bacterial strains. Moreover, Cu-substituted SBG-derived glass-ceramic was acknowledged for its excellent antimicrobial efficiency (≥99.9% reduction) against *P. aeruginosa*, *E. coli*, *Klebsiella aerogenes* and *S. aureus* [29]. Furthermore, Cu seems to have the ability to be a fungicide as shown in both silica- [26] and apatites-based materials [16]. The association of both Cu and Ga, as co-dopants in SBGs, can thereby be an attractive way to significantly boost and extend the antimicrobial range of such biomaterials. Furthermore, Cu and Ga have other biological benefits, strengthening their potential in healthcare. For example, Cu containing SBGs were reported to encourage the proliferation and differentiation of cells towards the osteogenic phenotype and inhibit osteoclast activity [14,30,31]. A similar behaviour has been reported for Ga ions [23,32,33].

Such results advocate for the investigation of the coupled effects of Cu and Ga in alkali-free SBG implant coatings. The benefits that could emerge from the future implementation of co-substituted SBG implant coatings, possessing both effective therapeutic response and adequate mechanical behaviour is nowadays a subject of intense research activity.

In this context, radio-frequency magnetron sputtering (RF-MS), a prominent physical vapour deposition technique, has been selected for the fabrication of implant coatings, due to its proficiency to synthesise high purity uniform films, and the potential to scale-up a refined process to industrial levels [5]. To the best of our knowledge, no studies have been performed yet on RF-MS synthesised SBG films containing Cu and/or Ga. As aforementioned, the poor interfacial bonding strength of typical SBG films may constitute their Achilles' heel in terms of their clinical applicability. The engineering of SBG film compositions by tuning the network formers/network modifiers ratio could be a strategy to follow when targeting mechanically performant SBG films. Previous attempts have shown that SBG RF-MS layers with a silica content in the range of ~55–60 mol.% presented excellent pull-off bonding strength values, exceeding 85 MPa [34], thus, being compliant to both recommendations of ISO 13779-2:2018 (minimum value of 15 MPa) [35] and FDA—STP1196:1994 draft guidance (minimum value of 50.8 MPa) [36] for implant coatings.

In this work, results concerning thin silica-based implant-type coatings with double addition of potentially antimicrobial Cu and Ga agents are presented for the first time, in terms of preliminary mechanical and in vitro biological behaviour. The study could be of importance for an accelerated development of SBG-based implant coatings with excellent mechanical properties, good cell viability and proliferation and high antimicrobial efficiency.

2. Materials and Methods

2.1. Synthesis of Source Cathode Target Materials

FastOs®BG (mol.%: SiO_2—38.49, CaO—36.07, P_2O_5—5.61, MgO—19.24, CaF_2—0.59) and Cu and Ga co-substituted FastOs®BG (mol.%: SiO_2—38.49, CaO—34.07, P_2O_5—5.61, MgO—16.24, CuO—2.00, Ga_2O_3—3.00, CaF_2—0.59) glasses were prepared by melt-quenching, following the protocol reported elsewhere [37,38]. The glass powders with fine granularity were obtained in a high-speed agate ball mill. The mean particle size of the SBG powders was of ~20–40 μm, as inferred from the particle size distribution curves obtained by the dynamic light scattering method (Beckman Coulter, model LS 230, Brea, CA, USA) by applying the Fraunhofer optical diffraction model.

Around 25 g of each SBG powder were mildly pressed (5 kgf) at room-temperature (RT) into copper dishes (diameter 110 mm, depth 3 mm) to obtain a magnetron cathode target. The benefits of using such type of targets (i.e., prevention of risk of target cracking and better control over the deposited coating composition) instead of classical ones, melt-poured into moulds, have been advocated in a previous work [39].

2.2. Deposition by RF-MS of Double-Layered Glass Coatings

SBG coatings were deposited onto 10 mm × 10 mm titanium (Mateck GmbH, Jülich, Germany) and silicon (Medapteh, Magurele, Romania) mirror polished substrates with a Vacma UVN-75-R1-type deposition system (Kazan, Soviet Union), equipped with planar magnetron cathodes (110 mm in diameter) and a RF generator of 1.78 MHz. The layer thicknesses were adjusted based on the deposition rates previously determined by spectroscopic ellipsometry.

The FastOs®BG and Cu and Ga substituted FastOs®BG layers with a thickness of ~600 nm were intentionally enriched in silica by applying the RF-MS protocol defined in Reference [40], with fused silica plates (Präzisions Glas & Optik GmbH, Iserlohn, Germany) positioned on the racetrack (annular zone of maximum sputtering erosion). All depositions were carried out at an argon pressure of 0.4 Pa and a target-to-substrate distance of 35 mm.

2.3. Physico-Chemical Characterization

The films' thickness was determined by spectroscopic ellipsometry (SE) using a Woollam apparatus (Lincoln, NE, USA) equipped with a HS-190 monochromator. The measurements were made at 65°, 70° and 75° incidence angles, with a resolution of 3°, in the 400–1700 nm spectral range. The SBG films were modelled using a Cauchy dispersion relation [41,42].

As their morphology was extremely smooth, the topological features of the SBG films were explored by atomic force microscopy (AFM), using an NT-MDT NTEGRA NanoLaboratory Probe system (Moscow, Russia). AFM images were recorded in non-contact mode on 5×5 μm^2 areas.

The elemental compositions of the source SBG materials and derived RF-MS coatings were estimated by energy dispersion X-ray spectroscopy (EDXS) using the EDAX Inc. (Mahwah, NJ, USA) micro-probe attached to a scanning electron microscope, operated at an acceleration voltage of 10 kV. The measurements were made on at least four randomly chosen sample regions with areas of 250×250 μm^2. Only the elemental (in at.%) concentrations of Si, Ca, P, Mg, Cu and Ga were inferred and will be further represented as arithmetic means ± standard deviations, since the quantification of lighter elements (i.e., oxygen and fluorine) by EDXS is prone to large errors [43].

The chemical state of SBG elements was analysed by X-ray photoelectron spectroscopy (XPS) using a Specs GmbH Multimethod System (Berlin, Germany). The XPS system is equipped with a Phoibos 150 hemispherical energy analyser with a multi-element two-stage transfer lens and a nine channeltron detector array. The photo-emission studies were carried out at a pressure of ~10^{-7} Pa, using a XR-50M Al Kα (1486.7 eV) source coupled to a FOCUS500 single-crystal quartz monochromator. The X-ray source was set at 300 W. A pass energy of 20 eV was used for the high-resolution core level spectral recordings. An ultimate resolution of 0.44 eV (given as full width at half maximum of the Ag $3d_{5/2}$ peak recorded with a pass energy of 20 eV, using monochromated Al Kα radiation, for a clean silver film) can be attained. The sample neutralization during the measurements was achieved by using a flood gun, working at acceleration energy of 1 eV and an emission current of 0.1 mA. In order to remove surface contaminants due to environmental/adventitious carbonaceous species, an in situ preliminary argon ions etching (using a Specs IQE11/35 ion gun) session was performed for 5 min, at an energy of 3 keV and a pressure of 1×10^{-3} Pa. The background was subtracted using the Shirley method. The fitting of spectra was performed with a dedicated software Spectral Data Processor using Voigt functions, version 3.0 (XPS International, Mountain View, CA, USA).

The structure of the SBG films was investigated by Fourier transform infrared (FTIR) spectroscopy measurements with a Perkin Elmer Spectrum BX II (Waltham, MA, USA) spectrophotometer. The FTIR spectroscopy investigations were performed in transmission mode, on films deposited onto infrared transparent Si substrates. The spectra were acquired in the wave numbers range of 4000–400 cm^{-1}, at a resolution of 4 cm^{-1}, and represent the average of 32 individual scans.

The surface free energy of the SBG films was determined based on static contact angle measurements (performed at least in triplicate), using a Drop Shape Analysis system (model DSA 100) from Krüss GmbH (Hamburg, Germany) and two standard solutions, one polar (i.e., water) and one dispersive (i.e., diiodomethane). The measurements were performed at RT, with the two solutions being precisely poured in droplets on the sample surface via a needle attached to a syringe pump controlled by the DSA3® software (Hamburg, Germany). The volume of the droplet (i.e., 1 μL) and the distance from the droplet to the sample was kept constant throughout the experiments. The contact angles were estimated, and the surface energy was calculated using the Owens–Wendt method [44].

The bonding strength of the films to the Ti substrate was estimated by the pull-off test method, using a DFD®Instruments PAThandy adhesion tester (Kristiansand, Norway), with a maximum force of 1 kN, and applying the testing procedure described in References [45,46], conducted in accordance with the ISO 4624:2016 [47] and ASTM D4541—17:2017 [48] standards.

The nano-mechanical properties were evaluated by nano-indentation tests using an NHT-2 CSM Instruments/Anton Paar GmbH (Peseux, Switzerland), equipped with a Berkovich diamond nano-indenter. The Oliver–Pharr method [49] was used to determine the modulus of elasticity (E) and hardness (H) of films from the load-displacement curves. The penetration depth was chosen in such a manner to minimize as much as possible the influence of the substrate on the results.

The scratch and wear resistances were evaluated with a dedicated MicroScratch Tester from CSM Instruments (Peseux, Switzerland), equipped with a diamond Rockwell conical indenter. The scratch test involves the passage (with a progressive applied load) of the Rockwell indenter, on the surface of the sample to result in a scratch. The load applied on the conical indenter increases progressively producing elastic and plastic deformations of the layer-substrate systems, which are becoming more and more pronounced. Conventionally, three critical load values are detected: (i) the load which causes the first deterioration of the cohesion of the layer (first cracks) (Lc1); (ii) the load leading to the partial removal of the coating (partial delamination) (Lc2); and (iii) the applied load that causes the removal of at least 50% of the deposited layer (Lc3) [50,51]. The device is equipped with sensors that measure and record the acoustic emissions during the scratch process, the penetration depth and the friction force. The adhesion to the substrate, as well as the cohesion of the SBG layers were evaluated according to the European standard EN1071-3/2005: "Advanced technical ceramics—Methods of test for ceramic coatings—Part 3: Determination of adhesion and other mechanical failure modes by a scratch test" [52] and the "Scratch test atlas of failure modes" guide [53]. The abrasive wear behaviour of the SBG films was determined using the same MicroScratch Tester, by using a heat-treated steel tip (Rockwell conical geometry) (HV = 1000) to employ a friction torque. While the sample is in translation motion, the frictional forces that occur between the sample and the tip are measured using a Linear Variable Differential Transformer sensor (CSM Instruments, Peseux, Switzerland). Five incremental load values (i.e., 1, 2, 3, 4 and 5 N) were used, with 10 passages per load at a travel speed of 4 mm/min. The wear track section areas for each sample and each applied load were assessed by transversal scanning with the same MicroScratch tester, using a 0.03 N applied load. The wear coefficient was calculated on the basis of the wear volume loss at the maximum used load (5 N), as function of the total test length.

2.4. *In Vitro Preliminary Biological Testing*

Prior to each in vitro test all samples were sterilized by applying a dry-heat procedure performed at 180 °C for 1 h [54,55].

2.4.1. Antibacterial Tests

The antimicrobial activity against the *Staphylococcus aureus* (ATCC® 6538) bacterial strain was performed according to the ISO 22196:2011 standard [56]. Briefly, suspensions of *S. aureus* in nutritional broth (Sanimed International Impex SRL, Bucharest, Romania) were prepared at a concentration of 10^6 CFU/mL. 50 μL of nutrient broth containing 5×10^3 viable CFUs were deposited on the surface of the specimens. Then, an inert sterile plastic film was placed over the samples and the plates were inserted into the incubator. After 24 h, 3 mL of nutrient broth was added to each well and the samples were completely scraped with a dedicated sterile rubber scrubber to detach any bacterial cells that could be encapsulated in the biofilm. 10-serial dilutions were prepared for each studied situation, and from each sample dilution two simple agar plates were seeded with 1 mL of the inoculum. After 24 h of incubation, the colonies on the plates were counted. The number of viable CFUs developed after 24 h in the presence of control and SBG coatings were calculated with the formula: 3 × dilution factor × (average number of colonies on plates).

2.4.2. Cytocompatibility Assays

The cytocompatibility tests were performed on mouse fibroblast cells NIH/3T3 (ATCC® CRL-1658™) in accordance with ISO 10993-5:2009 [57]. The protocol was refined in previous studies [58,59] and can be briefly summarised as follows:

Cell culture preparation: When cells reached confluence, they were detached with trypsin, collected and centrifuged at 200× g for 10 min after trypsin inactivation with soybean trypsin inhibitor. After centrifugation, the cells were re-suspended, and their number was adjusted to 10^5 cells/mL. On each sample, 10^4 cells were seeded in 100 µL medium. The plates were inserted into the incubator for cell adhesion. After 4 h, 400 µL of the Dulbecco's Modified Eagle Medium/Nutrient Mixture F-12 (DMEM-F12) (Sigma Aldrich, St. Louis, MO, USA) culture medium supplemented with 10% foetal bovine serum, was added. After 24 h, 50 µL of medium was collected to investigate the cytotoxic effect by the lactate dehydrogenase (LDH) assay (Thermo Scientific, Waltham, MA, USA).

Cell morphology: Epifluorescence microscopy was used to assess the morphology of 3T3 cells on the surface of the SBG samples. After 24 h of culture, the cells were fixed with 4% paraformaldehyde in phosphate buffered saline (PBS) for 15 min and then washed three times for 5 min with PBS. The cells were then incubated for 1 h at RT with 100 µL Phalloidin-AlexaFluor546 (Invitrogen, Carlsbad, CA, USA). After a series of three PBS washes of 10 min each, the cells were incubated with 1 µg/mL 4′, 6-diamidino-2-phenylindole (DAPI) (Sigma Aldrich, St. Louis, MO, USA). After a final washing step in PBS (two times for 10 min) and double distilled water (one time), the cells were imaged in a standard fluorescence microscope Leica DM6 B apparatus (Wetzlar, Germany), equipped with a Leica DFC 9000 GT camera and appropriate fluorescence objectives and filters.

Cell proliferation: The cellular proliferation was evaluated using the MTS (3-(4, 5-dimethyl thiazol-2-yl) 5-(3-carboxymethoxyphenyl)-2-(4-sulfophenyl)-2H-tetrazolium) kit (Promega Corporation, Madison, WI, USA). At the time of culturing, three wells were used to investigate the mitochondrial activity at T_0 moment. Three hours after seeding, the cells were analysed under an inverted microscope to observe their adhesion. The well was filled with 400 mL of fresh complete cell culture medium and then 80 µL MTS buffer was added. After 1 h in the incubator, 120 µL of medium was transferred to 96-wells plates and the absorbance was read at 490 nm using a Zenyth 3100 multimodal microplate reader (Anthos Labtec Instruments, Salzburg, Austria). The optical density (OD) at the T_0 seeding moment was calculated, and then was used to quantify the cell proliferation after 24 h.

Cell death: The cytotoxicity of SBG films was investigated using an LDH assay kit. After 24 h of cell culturing, 50 µL of the supernatant medium was collected and transferred into a 96-well microplate. An amount of 50 µL LDH substrate solution, prepared according to the manufacturer's instructions was added to each well. After 30 min of incubation, the reaction was stopped with 50 µL of stop buffer. The absorbance values at 490 and 620 nm were read with the Zenyth 3100 apparatus and the result was considered as the difference between the absorption at 490 nm and that at 620 nm (taken as a standard). The cut-off value was obtained in the presence of the fresh culture medium.

2.5. Statistical Analysis

The statistical analysis was performed using the unpaired two-tailed Student's t-test, with the differences being considered significant when $p < 0.05$.

3. Results and Discussion

3.1. Physico-Chemical Analyses

3.1.1. AFM Morphological Examination

The AFM microscopic investigations revealed that the SBG films, regardless of their composition (i.e., FastOs®BG—Figure 1a or Cu&Ga-FastOs®BG-derived—Figure 1b), had a highly similar morphology, being smooth (i.e., root-mean-square roughness of <1 nm) and compact. The films were composed of dome-shaped nano-sized particles with diameters within the ~15–20 nm narrow range, characteristic to amorphous magnetron sputtered structures.

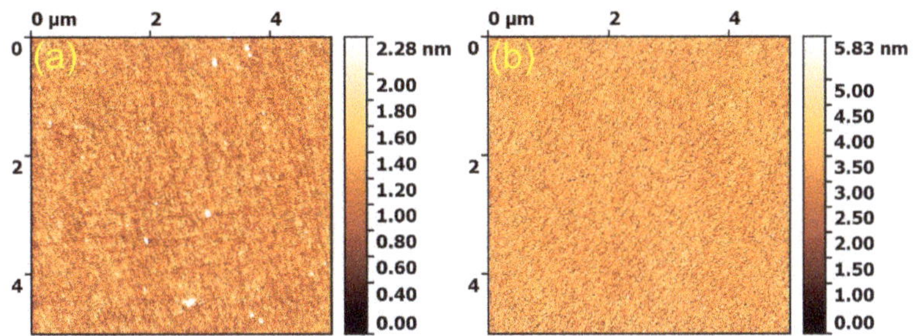

Figure 1. 2D atomic force microscopy (AFM) images recorded on 5×5 μm^2 surface areas of the (**a**) FastOs®BG- and (**b**) Cu&Ga-FastOs®BG-derived films.

3.1.2. EDXS Compositional Analysis

The comparative elemental compositions (in at.%) of the source SBG materials and of the FastOs®BG- and Cu&Ga-FastOs®BG-derived layers, as estimated by EDXS analysis, are presented in Figure 2. The results showed that the SBG films were, as intended, successfully enriched in Si/SiO_2 (~53–54 at.%/~56–57 mol.%) in both coating cases, at the expense of Ca and Mg, which recorded a concentration decrease of ~23–32% and ~18–24 at.%, respectively. The decrease (with 41–48 at.%) of the P concentration is a characteristic to RF-MS processes, and it is likely linked to the higher volatility of P_2O_5 (and associated with its relatively low sublimation latent heat) compared to the other oxide glass constituents [40,60]. The Cu and Ga contents of the Cu&Ga-FastOs®BG source biomaterial were well-transferred into the RF-MS films (no statistical significant differences being determined, $p > 0.05$).

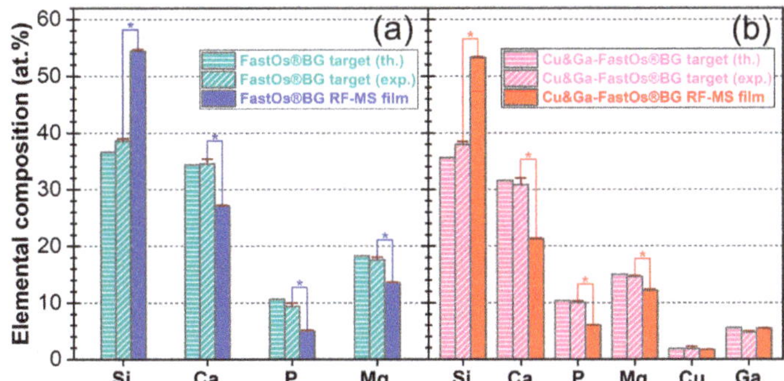

Figure 2. Comparative elemental composition (in at.%) of the magnetron cathode targets and the deposited silica-rich films: (**a**) FastOs®BG- and (**b**) Cu&Ga-FastOs®BG-based materials. The quantification of oxygen and fluorine was not performed because of the inaccuracy of EDXS analysis method with respect to light elements ($Z < 11$) [43]. * $p < 0.05$, statistically significant differences, as determined by using an unpaired two-tailed Student's *t*-test.

3.1.3. XPS Chemical State Examination

The high-resolution core electron XPS spectra of the Si 2p, Ca 2p, P 2p, Mg 2p, Cu $2p_{3/2}$, Ga $2p_{3/2}$ and O 1s levels of the source materials and derived RF-MS films are presented comparatively in Figure 3a–d, e–h, i–l, m–p, q–s, r–t, and u–x, respectively. The binding energy positions of the Si 2p (~101.9–102.6 eV), Ca $2p_{3/2}$ (~347.5–347.6 eV), P $2p_{3/2}$ (~133.0–133.4 eV), Mg 2p (~50.1–50.3 eV) correlated with the Auger parameter calculated as the binding energy of Mg 2p + kinetic energy of Mg KLL (~1230.0–1230.7 eV) lines (data not shown), disclosed their complete oxidation [61] in the case of both source materials and derived SBG films.

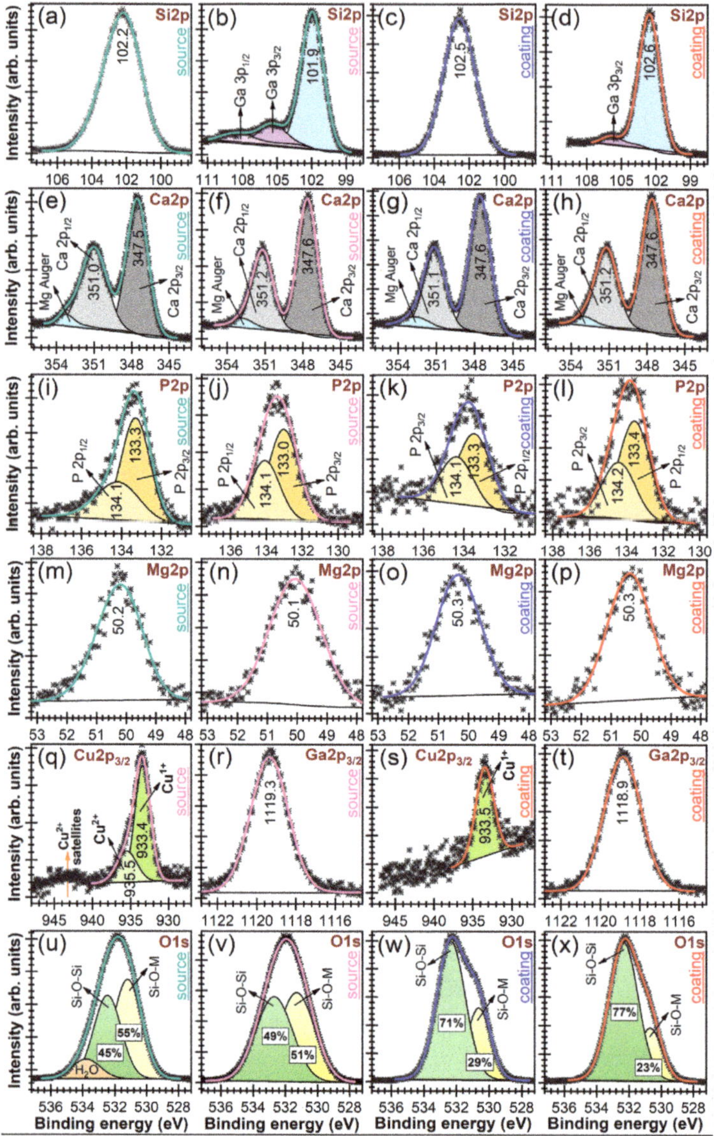

Figure 3. High resolution XPS spectra of the (**a–d**) Si 2p, (**e–h**) Ca 2p, (**i–l**) P 2p, (**m–p**) Mg 2p, (**q**,**s**) Cu $2p_{3/2}$, (**r**,**t**) Ga $3p_{3/2}$, (**u–x**) O 1s core photoelectron levels, recorded for the (**a**,**c**,**e**,**g**,**i**,**k**,**m**,**o**,**u**,**w**) FastOs®BG and (**b**,**d**,**f**,**h**,**j**,**l**,**n**,**p**,**q–t**,**v**,**x**) Cu&Ga-FastOs®BG (**a**,**b**,**e**,**f**,**i**,**j**,**m**,**n**,**q**,**r**,**u**,**v**) source materials and (**c**,**d**,**g**,**h**,**k**,**l**,**o**,**p**,**s**,**t**,**w**,**x**) derived films. Symbols: experimental spectrum, Solid line: fit sum. Filled area curves: spectral components.

The Ga $2p_{3/2}$ levels are situated at ~1119.3 and ~1118.9 eV for the cathode target and RF-MS film (Figure 3r,t), respectively, unveiling the 3^{+} oxidation state of gallium [61,62]. The core electron spectra collected in the region of the Cu $2p_{3/2}$ peak indicated that the Cu is found in both 1+ (~933.4 eV) and 2+ (~935.5 eV) oxidation states in the case of the Cu&Ga-FastOs®BG source material (Figure 3q) [61]. The presence of Cu^{2+} in the melt-quenched glass is also supported by the lower intensity broad peak, situated at higher binding energies and ascribed to the Cu^{2+} satellites [61]. In the case of the Cu&Ga-FastOs®BG-derived RF-MS films, the Cu^{2+} satellites were difficult to emphasize due to the low intensity of the peaks and consequently higher noise/signal ratio. The Cu^{2+} reduction to Cu^{+} oxidation state could be related to the lower thermodynamic driving force for the formation of CuO with respect to Cu_2O, as the Gibbs free energy of oxidation is higher for the former [63]. As a matter of fact, the Gibbs free energy of formation of the Cu oxides is lower with respect to all the other glass components (i.e., SiO_2, CaO, P_2O_5, MgO, Ga_2O_3) [63].

Two components were disclosed by the peak separation procedure applied to the O 1s spectra of the glass source materials and RF-MS films (Figure 3u–x). The higher energy component (positioned at ~532.3–532.8 eV) is assigned to the Si–O–Si bonds contribution (bridging oxygen bonds), whilst the lower binding energy one (centred at ~530.7–531.3 eV) is associated to the oxygen bonds with metal species (i.e., Si–O–M bonds (M = metal ions in the glass), non-bridging oxygen bonds) [64,65]. An additional third minor O 1s component, positioned at a higher binding energy (~533.8 eV), was evidenced in the case of the FastOs®BG target material only (Figure 3u). This third component can be ascribed to the adsorbed water [58,61], and is determined by hydroscopic character of the SBG. The Si–O–Si/Si–O–M ratio was dramatically increased in the case of RF-MS coatings with respect to the source materials (Figure 3u–x), as consequence of their intentional increase in Si concentration (Figure 2). In both the source material and RF-MS derived coating cases, the incorporation of Cu and Ga into the glass structure induced a slight increase of Si–O–Si/Si–O–M ratio, which suggested an improvement of the network connectivity.

3.1.4. FTIR Spectroscopy Structural Investigation

When network modifiers are incorporated into SBGs, the covalent Si–O–Si bonds are broken, and non-bridging oxygen atoms (NBO) are formed. The silica-based glass network becomes disrupted through the creation of ionic bonds between NBOs and modifier cations [5]. The network connectivity is directly influencing the stability and durability of SBG in contact with the intercellular fluids [5]. The short-range order in oxide glasses is generally quantified by the Q^n notation, in which Q represents a network-former polyhedron (in the case of SBG, $[SiO_4]$), whilst n corresponds to the number of associated bridging oxygen atoms (BOs).

The FTIR spectra of the simple and Cu & Ga substituted SBG source biomaterials, and of the RF-MS derived coatings are presented together in Figure 4a–c. Both the FastOs®BG-based source materials (Figure 4a) were characterized by the presence of large IR absorption bands with four maxima positioned at (i) ~1181 cm^{-1}, (ii) ~1030 cm^{-1}, (iii) ~950 cm^{-1} and (iv) 855 cm^{-1} appertaining to the asymmetric stretching (ν_{as}) vibrational modes of: (i + ii) of the Si–O–Si bonds in all the silicate tetrahedrons, TO_3 (Transverse-Optical) and LO_3 (Longitudinal-Optical) modes, respectively, and of Si–O bonds in (iii) $Q^2 + Q^3$ (with one and two NBOs) and (iv) $Q^0 + Q^1$ (with three and four NBOs) units [9,66–68]. The lower intensity band at ~749 cm^{-1} is assigned to the symmetrical stretching (ν_s) vibrations of Si–O bonds. The incorporation of Cu and Ga into the composition of the parent FastOs®BG material induced an improvement of the glass network connectivity, as suggested by the small intensity reduction of the $Q^2 + Q^3$ NBO band (at ~950 cm^{-1}), while the intensity of the Si–O–Si band (at ~1030 cm^{-1}) remained constant (Figure 4b).

The successful increase of the network modifiers concentration (Figure 2) of SBG RF-MS coatings had as consequences the (i) disappearance of the band at ~855 cm^{-1} (associated with the ν_{as} $Q^0 + Q^1$ band) (Figure 4c), together with a (ii) pronounced decrease in intensity of the ν_{as} $Q^2 + Q^3$ band (Figure 4b).

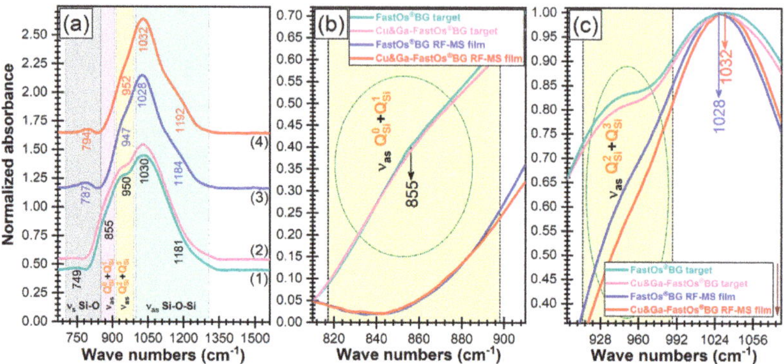

Figure 4. (**a**) FTIR spectra of the (1) FastOs®BG and (2) Cu&Ga-FastOs®BG melt-quenched glass powders, and (3) FastOs®BG- and (4) Cu&Ga-FastOs®BG-derived RF-MS layers; Insets highlighting the main structural modifications occurring in the spectral regions of the (**b**) ν_{as} Q^0 + Q^1 and (**c**) ν_{as} Q^2 + Q^3 vibration bands.

The incorporation of Cu & Ga therapeutic ions into the SBF RF-MS film structure led to minor structural changes that advocate for a slight polymerization of this type of coating with respect to the FastOs®BG-derived one (similarly to the trends experienced by the source materials). This was hinted by the minor intensity decrease and position shift of the absorption band associated with ν_{as} vibration modes of bonds in the Q^3 + Q^2 units (Figure 4b). Concurrently, the IR absorption bands generated by Si–O–Si bonds experienced a slight blue-shift (Figure 4a,b). This is in agreement with the afore-presented XPS results (Figure 3u–x), which indicated a higher Si–O–Si/Si–O–M ratio in the case of the Cu&Ga-FastOs®BG-based materials. Since the network former content (Si + P) is the same (~59.3%) for both type of films, and the concentration of well-known network modifiers (Ca + Mg + Cu) is higher (~40.6% vs. ~35.2%) for the FastOs®BG-derived structure (Figure 2), it is suggested that at least part of Ga ions act as network formers. The network former role of Ga is not unprecedented [23] in SBGs. However, for the time being this should be treated as merely a hypothesis, with the clarification of Ga role within the RF-MS SBG films' structure demanding further insightful analyses, which will be the focus of future systematic studies.

3.1.5. Surface Energy Measurements

The evolution of surface free energy (and their polar and dispersive components as nominal values and ratios) and water contact angles (CA), recorded for the bare Ti and simple and Cu-Ga doped SBG coatings are shown in Figure 5a,b. The variation of the polar and dispersive components of the surface free energy (SFE, γ_{tot}), and thus the hydrophilic/hydrophobic character, can unveil prospective biofunctional traits of a scrutinized material/construct [69,70]. The polar component (γ_p) is generated by the chemical bonds/interactions (e.g., dipole-dipole interactions) within the material, whilst the dispersive component (γ_d) is linked to the movement of electrons around atoms/molecules and temporary variation in the electron density with associated temporary dipoles [71]. SFE is known to play a major role in biocompatibility, leading to the arrangement of functional groups and electrical charges on the surface of the biomaterials in contact with the living environment, and thus, govern the first interactions with the intercellular fluid and the adherence of cells [70,72,73].

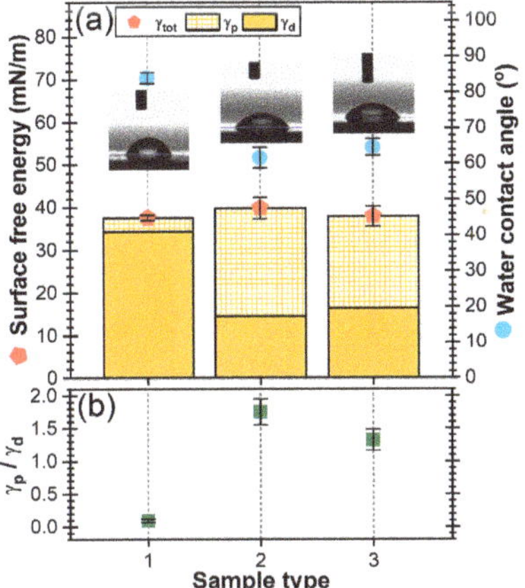

Figure 5. Evolution of the (**a**) surface free energy values (γ_{tot}), its polar (γ_p) and dispersive (γ_d) components, water contact angle average values and (**b**) ratios of the polar (γ_p)/dispersive (γ_d) components of the surface free energy, for the (1) bare Ti and (2) FastOs®BG-derived and (3) Cu&Ga-FastOs®BG-derived RF-MS coatings.

The SFE values of both the bare and FastOs®BG- and Cu&Ga-FastOs®BG-derived coated Ti specimens were found to be situated in a rather narrow range of ~37–40 mN/m ($p > 0.05$, thus, without statistical significant differences) (Figure 5a). However, the weight of the polar and dispersive components of the SFE significantly differed in the case of the silica-rich SBG coated Ti, with the polar component becoming dominant (Figure 5b). While the contact angle (CA) value with water does not change upon the incorporation of Cu & Ga into the SBG film structure (CA ≈ 62–64°, $p > 0.05$), in the case of Ti, a CA value towards the hydrophobic domain (i.e., 84°) was obtained. It is important to note that in the case of implant surfaces, many studies indicated that an optimal wettability range, capable of augmenting the cellular response (i.e., adhesion, proliferation and cytoskeleton organisation), is the 60–80° one [73–75]. Thereby, from this point of view favourable premises existed for a positive biological response in the case of the proposed silica-rich bioglass implant coating design.

3.2. Mechanical Performance Characterization

The first mechanical assessment consisted in the evaluation of the bonding strength of the SBG-based coating by the pull-off test. The tensile tests yielded similar values for both simple and Cu-Ga doped SBG coatings (Figure 6a), with average bonding strength values of ~53–55 MPa ($p > 0.05$, thus, without statistically significant differences). The recorded bonding strength values are situated above the limits imposed for implant coatings by the ISO 13779–2:2018 [35] (i.e., 15 MPa), and are close to those recommended by the FDA—STP1196 draft guidance (i.e., 50.8 MPa) [36].

Further, the mechanical properties of the SBG films were evaluated by scratch, nano-indentation and wear tests.

During the scratch tests, the Lc1 load value (the first indication of cohesive failure (appearance of cracks)) was not observed, while the Lc2 (load responsible for the first delamination) and Lc3 (load at which the coating is severely delaminated from the substrate) were clearly observed on the scratch tracks. The Lc2 and Lc3 values of FastOs®BG- and Cu&Ga-FastOs®BG-derived RF-MS coatings are presented in Figure 6b. The incorporation of Cu and Ga into the silica-rich SBG RF-MS layer led to a statistically significant ($p < 0.05$) increase of ~15%, with the Lc3 having a value of 4.9 ± 0.47 N. These values are similar to those (i.e., ~5 N) required to partially delaminate bioglass coatings of similar thickness, deposited by pulsed electron deposition (PED) from a silica-rich system (mol.%:

SiO_2–47.2, Ca–45.6, P_2O_5–2.6, K_2O–4.6), and superior to those (~2 N) of 600 °C heat-treated 45S5 PED coatings [76].

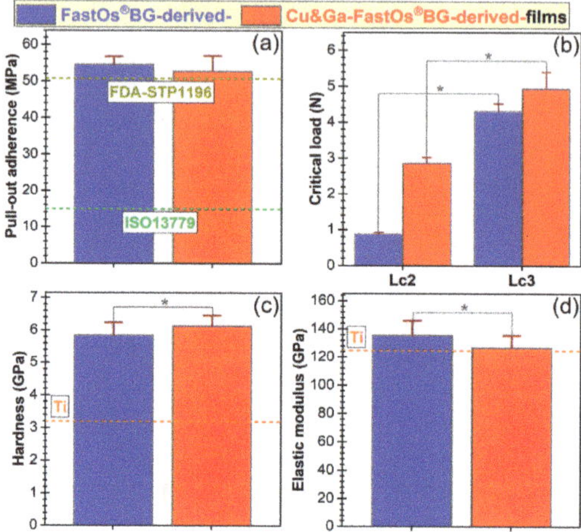

Figure 6. Summary of the mechanical performance of the SBG films in terms of (**a**) pull-off bonding strength (tensile), (**b**) scratch resistance, (**c**) hardness and (**d**) elastic modulus. * $p < 0.05$, statistically significant differences, as determined by using an unpaired two-tailed Student's *t*-test.

The average hardness and elastic modulus values of the SBG coatings, determined by nano-indentation measurements, are presented in Figure 6c,d, respectively. The Cu & Ga doping moderately increased ($p < 0.05$) the hardness of the RF-MS film to a value of ~6.1 GPa (Figure 6c), whilst the elastic modulus slightly decreased ($p < 0.05$) to ~127 GPa (Figure 6d). The nano-mechanical performances were similar (E) or even higher (H) with respect to titanium (e.g., E ≈ 120 GPa, H ≈ 3.6 GPa [58,77]).

Furthermore, the wear behaviour against a steel Rockwell tip, in linear motion, was assessed using the same MicroScratch tester. The wear tracks obtained on the uncoated (Figure 7a) and FastOs®BG-coated (Figure 7b) Ti substrate were severe, regardless of the applied load size (1–5 N), with increasing widths and depths of the wear track for higher applied loads. In contrast, the Cu&Ga-FastOs®BG-derived coating showed a significantly enhanced wear behaviour (Figure 7c). For the lowest applied loads, namely 1, 2 and 3 N, some wear grooves are visible on the surface, but to a much smaller extent compared to the uncoated and FastOs®BG-coated Ti substrate. Moreover, the width of the wear tracks on the Cu&Ga-FastOs®BG coated sample is significantly smaller. This is a good indicator of an improved wear behaviour. No signs of delamination are visible for these loads. For higher applied loads, namely 4 and 5 N, coating failures (delamination) are visible on the wear tracks, after 10 reciprocating passes. These values are in good agreement with the ones obtained during the adhesion tests. The wear coefficients determined on the basis of the wear volume loss under a load of 5 N as a function of the total test length are presented in Figure 7d. As hinted by the optical microscopy images, the wear coefficient is significantly improved by the application of the Cu&Ga-FastOs®BG-derived coating. The so-called elastic strain to failure ratio (H/E) is a reliable indicator of wear resistance of materials, with values higher than 0.1 usually indicating a tough coating. Even though this value is not reached by the coatings presented herein, a slightly higher value of this ratio is exhibited by the Cu&Ga-FastOs®BG-derived coating, in good agreement with the other wear parameters (wear coefficient, wear track width). These results are quite promising at this stage of the research development, especially considering that bio-glasses are relatively soft.

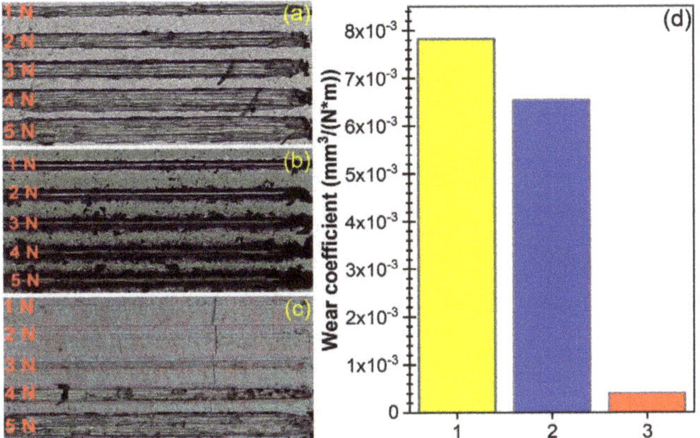

Figure 7. Optical microscopy images of the wear tracks recorded at five incremental load values (i.e., 1, 2, 3, 4 and 5 N), with 10 passes per load at a displacement rate of 4 mm/min, in the case of the (**a**) bare and (**b**) FastOs®BG- and (**c**) Cu&Ga-FastOs®BG-coated Ti substrate. (**d**) The wear coefficient determined for the three studied specimens under the load of 5 N.

3.3. *In Vitro Preliminary Biological Evaluation*

3.3.1. Antibacterial Efficacy at 24 h

Since dental implant coatings are the main targeted application, the antibacterial properties of the SBG films were preliminarily assessed against the Gram-positive *S. aureus* (ATCC® 6538) bacterial strain, both because of its presence in oral microbiota, and according to the ISO 22196:2011 recommendation [56]. 50,000 CFUs were seeded onto the samples and allowed to grow for 24 h. The silica-rich FastOs®BG did not impede the bacterial proliferation, whilst the Cu&Ga-FastOs®BG layer was found to reduce the bacterial development by 30 times with respect to the seeding CFU, and with 4 orders of magnitude with respect to the control situations (i.e., nutritive broth and bare substrate) (Figure 8). This output could be considered encouraging for combating the microbial infection at the implantation site. However, future dynamic studies need to be carried out, at several time intervals, and against a large palette of microbial strains, in order to fully probe and unveil the complete potential of such implant-type coatings.

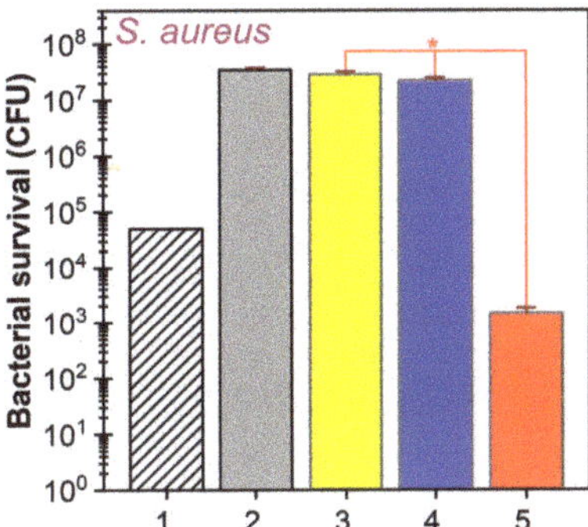

Figure 8. Antibacterial activity against *S. aureus* at 24 h. The data are presented in logarithmic values of CFUs (bacterial survival). (1) seeded CFUs; (2) control = nutrient broth without sample; (3) bare and silica-rich (4) FastOs®BG and (5) Cu&Ga-FastOs®BG coated substrate. * $p < 0.05$, statistically significant differences, as determined by using an unpaired two-tailed Student's *t*-test.

3.3.2. Cytocompatibility Response at 24 h

The cytocompatibility of the samples was tested on NIH/3T3 fibroblast cell cultures (ATCC® CRL-1658™). The cell proliferation was assessed by the MTS assay (Figure 9a), while the cell death was evaluated by a LDH test. Both silica-rich FastOs®BG and Cu&Ga-FastOs®BG coatings elicited an excellent proliferation of cells, with values similar to those recorded on the standard substrate for cell cultures (i.e., polycarbonate for cell cultures) (Figure 9a). Furthermore, the Cu&Ga-FastOs®BG-derived coatings showed lower values of LDH activity with respect to the FastOs®BG-derived ones, being situated closer to the ones recorded for the polycarbonate surface for cell cultures (Figure 9b). Thereby, it is advocated that the introduction of low concentrations of Cu and Ga into the thin FastOs®BG-derived films does not alter their cytocompatibility. When cultured on the silica-rich FastOs®BG- and Cu&Ga-FastOs®BG-derived coatings, the cells retained their characteristic morphology as seen in Figure 9c,d, respectively. Actin filaments grouped in bundles spanned the cells in a usual manner, whilst the nuclei retained their normal shape and chromatin condensation pattern.

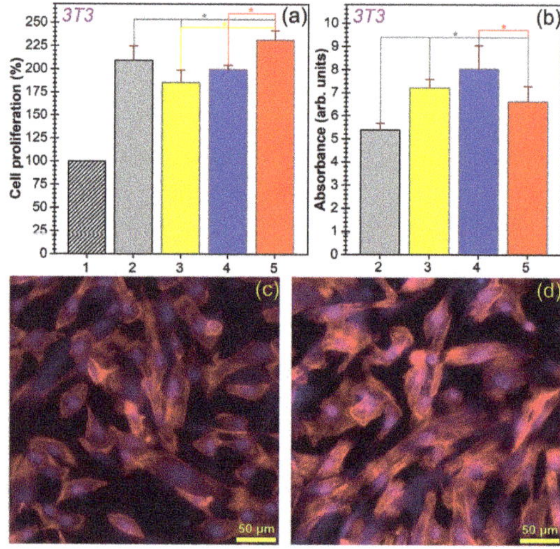

Figure 9. (**a**) Cell viability/proliferation assessed by MTS assay and (**b**) cytotoxicity assessed by LDH release after 24 h. (1) seeded cells; (2) biological control—polycarbonate; (3) bare and silica-rich (4) FastOs®BG and (5) Cu&Ga-FastOs®BG coated substrate. * $p < 0.05$, statistically significant differences, as determined by using an unpaired two-tailed Student's *t*-test. Epi-fluorescence microscopy images revealing the morphology of 3T3 fibroblast cells grown on the silica-rich: (**c**) FastOs®BG and (**d**) Cu&Ga-FastOs®BG films. The actin cytoskeleton was stained with phalloidin-AlexaFluor596 (red), whilst cell nuclei were counterstained with DAPI (blue). Objective: 40×. Magnification bar: 50 μm.

When analysing the antibacterial activity and cytocompatibility responses together, it is suggested that the application of layers of Cu&Ga-FastOs®BG on metallic endo-osseous implants represents a promising conceptual solution to minimize the risk of post-surgical bacterial infection.

4. Conclusions

Thin films of FastOs®BG alkali-free bioactive films co-doped with copper and gallium were successfully synthesised by radio-frequency magnetron sputtering.

The introduction of Cu and Ga induced a slight polymerization of the silica glass network of the films with respect to the undoped ones, having as effect a slight increase of the hardness (from ~5.8 to ~6.1 GPa) and critical load of scratch delamination (from ~4.3 to ~4.9 N), and a decrease of the elastic modulus (from ~136 to 127 GPa), with the bonding strength being conserved (~54 MPa).

Furthermore, the Cu&Ga-FastOs®BG-derived coating showed a significantly improved wear behaviour with respect to bare titanium and to the FastOs®BG coating.

While the contact angle with water is kept within the optimal 60–80° range for cell adhesion, and the cytocompatibility of the Cu & Ga silica-rich films remained unaltered, they manifested a marked antibacterial effect against the *S. aureus* strain, reducing its development by ~4 orders of magnitude after 24 h.

Overall, these preliminarily mechanical and in vitro biological performances of Cu & Ga co-substituted silica-based bioactive glass films are testifying for their certain promise, which demands further exploration, on route to bio-functionalisation solutions capable to protect metallic endo-osseous implants against post-surgical microbial infection.

Author Contributions: Conceptualization, G.E.S., A.-C.P., C.B., T.T., and J.M.F.F.; Powder synthesis: H.R.F. and J.M.F.F.; Film fabrication: G.E.S., T.T., I.M.C., and A.C.S.; AFM, FTIR spectroscopy analyses: T.T., I.M.C., C.B., A.C.S., and G.E.S.; XPS investigations: C.C.N.; Surface energy investigations: I.Z.; EDXS measurements: G.P.-P.; Mechanical tests: D.C.; Biological assays: A.-C.P., L.E.I., and M.N.; writing—original draft preparation: G.E.S. and T.T.; writing—review and editing: T.T., G.E.S., A.-C.P., and J.M.F.F.; funding acquisition: G.E.S. All authors have read and agreed to the published version of the manuscript.

Funding: This research was funded by Romanian National Authority for Scientific Research and Innovation (CNCS-UEFISCDI) in the framework of project PN-III-P1-1.1-TE-2016-1501. J.M.F.F. is thankful for the financial support from CICECO—Aveiro Institute of Materials and FCT Ref. UID/CTM/50011/2019 grant, financed through the FCT/MCTES.

Acknowledgments: The authors thank for the financial support of the Romanian National Authority for Scientific Research and Innovation (CNCS-UEFISCDI) in the framework of project PN-III-P1-1.1-TE-2016-1501, and to the institutional Core Program 21N. This work was also developed within the scope of the project CICECO-Aveiro Institute of Materials, FCT Ref. UID/CTM/50011/2019, financed by national funds through the FCT/MCTES. NIMP authors acknowledge with thanks the acquisition of the Leica DM6 B fluorescence microscope in the framework of the Operational Programme Competitiveness project NANOBIOSURF-SMIS 103528 (2014–2020). D.C. acknowledges the structural funds project PRO-DD (POS-CCE, O.2.2.1., ID123, SMIS 2637, ctr. no 11/2009) for providing the CSM Instruments infrastructure used in this work.

Conflicts of Interest: The authors declare no conflict of interest.

References

1. Kawaguchi, K.; Iijima, M.; Endo, K.; Mizoguchi, I. Electrophoretic deposition as a new bioactive glass coating process for orthodontic stainless steel. *Coatings* **2017**, *7*, 199. [CrossRef]
2. Baino, F.; Hamzehlou, S.; Kargozar, S. Bioactive glasses: Where are we and where are we going? *J. Funct. Biomater.* **2018**, *9*, 25. [CrossRef] [PubMed]
3. Brunello, G.; Elsayed, H.; Biasetto, L. Bioactive glass and silicate-based ceramic coatings on metallic implants: Open challenge or outdated topic? *Materials* **2019**, *12*, 2929. [CrossRef] [PubMed]
4. Leite, Á.J.; Gonçalves, A.I.; Rodrigues, M.T.; Gomes, M.E.; Mano, J.F. Strontium-doped bioactive glass nanoparticles in osteogenic commitment. *ACS Appl. Mater. Interfaces* **2018**, *10*, 23311–23320. [CrossRef] [PubMed]
5. Fernandes, H.R.; Gaddam, A.; Rebelo, A.; Brazete, D.; Stan, G.E.; Ferreira, J.M.F. Bioactive glasses and glass-ceramics for healthcare applications in bone regeneration and tissue engineering. *Materials* **2018**, *11*, 2530. [CrossRef] [PubMed]
6. Prefac, G.A.; Milea, M.L.; Vadureanu, A.M.; Muraru, S.; Dobrin, D.I.; Isopencu, G.O.; Jinga, S.I.; Raileanu, M.; Bacalum, M.; Busuioc, C. CeO_2 containing thin films as bioactive coatings for orthopaedic implants. *Coatings* **2020**, *10*, 642. [CrossRef]
7. Sergi, R.; Bellucci, D.; Cannillo, V. A comprehensive review of bioactive glass coatings: State of the art, challenges and future perspectives. *Coatings* **2020**, *10*, 757. [CrossRef]
8. Hench, L.L. The story of Bioglass®. *J. Mater. Sci. Mater. Med.* **2006**, *17*, 967–978. [CrossRef]
9. Popa, A.C.; Fernandes, H.R.; Necsulescu, M.; Luculescu, C.; Cioangher, M.; Dumitru, V.; Stuart, B.W.; Grant, D.M.; Ferreira, J.M.F.; Stan, G.E. Antibacterial efficiency of alkali-free bio-glasses incorporating ZnO and/or SrO as therapeutic agents. *Ceram. Int.* **2019**, *45*, 4368–4380. [CrossRef]

10. Stan, G.E.; Popescu, A.C.; Mihailescu, I.N.; Marcov, D.A.; Mustata, R.C.; Sima, L.E.; Petrescu, S.M.; Ianculescu, A.; Trusca, R.; Morosanu, C.O. On the bioactivity of adherent bioglass thin films synthesized by magnetron sputtering techniques. *Thin Solid Film.* **2010**, *518*, 5955–5964. [CrossRef]
11. Ciraldo, F.E.; Boccardi, E.; Melli, V.; Westhauser, F.; Boccaccini, A.R. Tackling bioactive glass excessive in vitro bioreactivity: Preconditioning approaches for cell culture tests. *Acta Biomater.* **2018**, *75*, 3–10. [CrossRef] [PubMed]
12. Bakry, A.S.; Tamura, Y.; Otsuki, M.; Kasugai, S.; Ohya, K.; Tagami, J. Cytotoxicity of 45S5 bioglass paste used for dentine hypersensitivity treatment. *J. Dent.* **2011**, *39*, 599–603. [CrossRef] [PubMed]
13. Ferreira, J.M.F.; Rebelo, A. The key Features expected from a perfect bioactive glass—How Far we still are from an ideal composition? *Biomed. J. Sci. Tech. Res.* **2017**, *1*, 936–939. [CrossRef]
14. Cacciotti, I. Bivalent cationic ions doped bioactive glasses: The influence of magnesium, zinc, strontium and copper on the physical and biological properties. *J. Mater. Sci.* **2017**, *52*, 8812–8831. [CrossRef]
15. Kaya, S.; Cresswell, M.; Boccaccini, A.R. Mesoporous silica-based bioactive glasses for antibiotic-free antibacterial applications. *Mater. Sci. Eng. C* **2018**, *83*, 99–107. [CrossRef]
16. Tite, T.; Popa, A.C.; Balescu, L.M.; Bogdan, I.M.; Pasuk, I.; Ferreira, J.M.F.; Stan, G.E. Cationic substitutions in hydroxyapatite: Current status of the derived biofunctional effects and their in vitro interrogation methods. *Materials* **2018**, *11*, 2081. [CrossRef]
17. Drago, L.; Toscano, M.; Bottagisio, M. Recent evidence on bioactive glass antimicrobial and antibiofilm activity: A mini-review. *Materials* **2018**, *11*, 326. [CrossRef]
18. Jones, J.R. Review of bioactive glass: From Hench to hybrids. *Acta Biomater.* **2013**, *9*, 4457–4486. [CrossRef]
19. Valappil, S.P.; Ready, D.; Abou Neel, E.A.; Pickup, D.M.; Chrzanowski, W.; O'Dell, L.A.; Newport, R.J.; Smith, M.E.; Wilson, M.; Knowles, J.C. Antimicrobial gallium-doped phosphate-based glasses. *Adv. Funct. Mater.* **2008**, *18*, 732–741. [CrossRef]
20. Valappil, S.P.; Ready, D.; Abou Neel, E.A.; Pickup, D.M.; O'Dell, L.A.; Chrzanowski, W.; Pratten, J.; Newport, R.J.; Smith, M.E.; Wilson, M.; et al. Controlled delivery of antimicrobial gallium ions from phosphate-based glasses. *Acta Biomater.* **2009**, *5*, 1198–1210. [CrossRef]
21. Gross, T.M.; Lahiri, J.; Golas, A.; Luo, J.; Verrier, F.; Kurzejewski, J.L.; Baker, D.E.; Wang, J.; Novak, P.F.; Snyder, M.J. Copper-containing glass ceramic with high antimicrobial efficacy. *Nat. Commun.* **2019**, *10*, 1979. [CrossRef] [PubMed]
22. Pourshahrestani, S.; Zeimaran, E.; Kadri, N.A.; Gargiulo, N.; Samuel, S.; Naveen, S.V.; Kamarul, T.; Towler, M.R. Gallium-containing mesoporous bioactive glass with potent hemostatic activity and antibacterial efficacy. *J. Mater. Chem. B* **2016**, *4*, 71–86. [CrossRef] [PubMed]
23. Sanchez-Salcedo, S.; Malavasi, G.; Salinas, A.J.; Lusvardi, G.; Rigamonti, L.; Menabue, L.; Vallet-Regi, M. Highly-bioreactive silica-based mesoporous bioactive glasses enriched with gallium(III). *Materials* **2018**, *11*, 367. [CrossRef] [PubMed]
24. Wu, C.; Zhou, Y.; Xu, M.; Han, P.; Chen, L.; Chang, J.; Xiao, Y. Copper-containing mesoporous bioactive glass scaffolds with multifunctional properties of angiogenesis capacity, osteostimulation and antibacterial activity. *Biomaterials* **2013**, *34*, 422–433. [CrossRef]
25. Bari, A.; Bloise, N.; Fiorilli, S.; Novajra, G.; Vallet-Regí, M.; Bruni, G.; Torres-Pardo, A.; González-Calbet, J.M.; Visai, L.; Vitale-Brovarone, C. Copper-containing mesoporous bioactive glass nanoparticles as multifunctional agent for bone regeneration. *Acta Biomater.* **2017**, *55*, 493–504. [CrossRef]
26. Peszke, J.; Dulski, M.; Nowak, A.; Balin, K.; Zubko, M.; Sułowicz, S.; Nowak, B.; Piotrowska-Seget, Z.; Talik, E.; Wojtyniak, M.; et al. Unique properties of silver and copper silica-based nanocomposites as antimicrobial agents. *RSC Adv.* **2017**, *7*, 28092–28104. [CrossRef]
27. Ryan, E.J.; Ryan, A.J.; González-Vázquez, A.; Philippart, A.; Ciraldo, F.E.; Hobbs, C.; Nicolosi, V.; Boccaccini, A.R.; Kearney, C.J.; O'Brien, F.J. Collagen scaffolds functionalised with copper-eluting bioactive glass reduce infection and enhance osteogenesis and angiogenesis both in vitro and in vivo. *Biomaterials* **2019**, *197*, 405–416. [CrossRef]
28. Rau, J.V.; Curcio, M.; Raucci, M.G.; Barbaro, K.; Fasolino, I.; Teghil, R.; Ambrosio, L.; De Bonis, A.; Boccaccini, A.R. Cu-releasing bioactive glass coatings and their in vitro properties. *ACS Appl. Mater. Interfaces* **2019**, *11*, 5812–5820. [CrossRef]
29. Miola, M.; Verné, E. Bioactive and antibacterial glass powders doped with copper by ion-exchange in aqueous solutions. *Materials* **2016**, *9*, 405. [CrossRef]

30. Hoppe, A.; Meszaros, R.; Stähli, C.; Romeis, S.; Schmidt, J.; Peukert, W.; Marelli, B.; Nazhat, S.N.; Wondraczek, L.; Lao, J.; et al. In vitro reactivity of Cu doped 45S5 Bioglass® derived scaffolds for bone tissue engineering. *J. Mater. Chem. B* **2013**, *1*, 5659–5674. [CrossRef]
31. Romero-Sánchez, L.B.; Marí-Beffa, M.; Carrillo, P.; Medina, M.Á.; Díaz-Cuenca, A. Copper-containing mesoporous bioactive glass promotes angiogenesis in an in vivo zebrafish model. *Acta Biomater.* **2018**, *68*, 272–285. [CrossRef] [PubMed]
32. Yu, M.; Wang, Y.; Zhang, Y.; Cui, D.; Gu, G.; Zhao, D. Gallium ions promote osteoinduction of human and mouse osteoblasts via the TRPM7/Akt signaling pathway. *Mol. Med. Rep.* **2020**, *22*, 2741–2752. [CrossRef] [PubMed]
33. Wajda, A.; Goldmann, W.H.; Detsch, R.; Grünewald, A.; Boccaccini, A.R.; Sitarz, M. Structural characterization and evaluation of antibacterial and angiogenic potential of gallium-containing melt-derived and gel-derived glasses from CaO-SiO_2 system. *Ceram. Int.* **2018**, *44*, 22698–22709. [CrossRef]
34. Stan, G.E.; Morosanu, C.O.; Marcov, D.A.; Pasuk, I.; Miculescu, F.; Reumont, G. Effect of annealing upon the structure and adhesion properties of sputtered bio-glass/titanium coatings. *Appl. Surf. Sci.* **2009**, *255*, 9132–9138. [CrossRef]
35. ISO—International Organization for Standardization. *ISO 13779-2:2018—Implants for Surgery—Hydroxyapatite—Part 2: Thermally Sprayed Coatings of Hydroxyapatite*; ISO—International Organization for Standardization: Geneva, Switzerland, 2018.
36. Callahan, T.J.; Gatenberg, J.B.; Sands, B.E. STP1196—Calcium phosphate (Ca-P) coating draft guidance for preparation of food and drug administration (FDA) submissions for orthopedic and dental endosseous implants. In *Characterization and Performance of Calcium Phosphate Coatings for Implants*; ASTM International: West Conshohocken, PA, USA, 1994; pp. 185–197.
37. Goel, A.; Kapoor, S.; Rajagopal, R.R.; Pascual, M.J.; Kim, H.-W.; Ferreira, J.M.F. Alkali-free bioactive glasses for bone tissue engineering: A preliminary investigation. *Acta Biomater.* **2012**, *8*, 361–372. [CrossRef]
38. Kapoor, S.; Goel, A.; Tilocca, A.; Dhuna, V.; Bhatia, G.; Dhuna, K.; Ferreira, J.M.F. Role of glass structure in defining the chemical dissolution behavior, bioactivity and antioxidant properties of zinc and strontium co-doped alkali-free phosphosilicate glasses. *Acta Biomater.* **2014**, *10*, 3264–3278. [CrossRef]
39. Popa, A.C.; Stan, G.E.; Besleaga, C.; Ion, L.; Maraloiu, V.A.; Tulyaganov, D.U.; Ferreira, J.M.F. Submicrometer hollow bioglass cones deposited by radio frequency magnetron sputtering: Formation mechanism, properties, and prospective biomedical applications. *ACS Appl. Mater. Interfaces* **2016**, *8*, 4357–4367. [CrossRef]
40. Popa, A.C.; Stan, G.E.; Husanu, M.A.; Mercioniu, I.; Santos, L.F.; Fernandes, H.R.; Ferreira, J.M.F. Bioglass implant-coating interactions in synthetic physiological fluids with varying degrees of biomimicry. *Int. J. Nanomed.* **2017**, *12*, 683–707. [CrossRef]
41. Cauchy, A.L. Sur la réfraction et la réflexion de la lumière. *Bull. des Sci. Math. (Bull. Férussac)* **1830**, *14*, 6–10.
42. Galca, A.C.; Stan, G.E.; Trinca, L.M.; Negrila, C.C.; Nistor, L.C. Structural and optical properties of c-axis oriented aluminum nitride thin films prepared at low temperature by reactive radio-frequency magnetron sputtering. *Thin Solid Film.* **2012**, *524*, 328–333. [CrossRef]
43. ISO—International Organization for Standardization. *ISO 22309:2011—Microbeam Analysis—Quantitative Analysis Using Energy-Dispersive Spectrometry (EDS) for Elements with an Atomic Number of 11 (Na) or Above*; ISO—International Organization for Standardization: Geneva, Switzerland, 2011.
44. Owens, D.K.; Wendt, R.C. Estimation of the surface free energy of polymers. *J. Appl. Polym. Sci.* **1969**, *13*, 1741–1747. [CrossRef]
45. Popa, A.C.; Stan, G.E.; Husanu, M.A.; Pasuk, I.; Popescu, I.D.; Popescu, A.C.; Mihailescu, I.N. Multi-layer haemocompatible diamond-like carbon coatings obtained by combined radio frequency plasma enhanced chemical vapor deposition and magnetron sputtering. *J. Mater. Sci. Mater. Med.* **2013**, *24*, 2695–2707. [CrossRef] [PubMed]
46. Visan, A.; Grossin, D.; Stefan, N.; Duta, L.; Miroiu, F.M.; Stan, G.E.; Sopronyi, M.; Luculescu, C.; Freche, M.; Marsan, O.; et al. Biomimetic nanocrystalline apatite coatings synthesized by matrix assisted pulsed laser evaporation for medical applications. *Mater. Sci. Eng. B* **2014**, *181*, 56–63. [CrossRef]
47. ISO—International Organization for Standardization. *ISO 4624:2016—Paints and Varnishes—Pull-off Test for Adhesion*; ISO—International Organization for Standardization: Geneva, Switzerland, 2016.
48. ASTM International. *ASTM D4541-17 Standard Test Method for Pull-Off Strength of Coatings Using Portable Adhesion*; ASTM International: West Conshohocken, PA, USA, 2017; pp. 1–16.

49. Oliver, W.C.; Pharr, G.M. An improved technique for determining hardness and elastic modulus using load and displacement sensing indentation experiments. *J. Mater. Res.* **1992**, *7*, 1564–1583. [CrossRef]
50. Jacobs, R.; Meneve, J.; Dyson, G.; Teer, D.G.; Jennett, N.M.; Harris, P.; von Stebut, J.; Comte, C.; Feuchter, P.; Cavaleiro, A.; et al. A certified reference material for the scratch test. *Surf. Coat. Technol.* **2003**, *174–175*, 1008–1013. [CrossRef]
51. Ingelbrecht, C.; Jennett, N.; Jacobs, R.; Meneve, J. *The Certification of Critical Coating Failure Loads: A Reference Material for Scratch Testing According to ENV 1071-3*; European Commission: Luxembourg, 1994; ISBN 928946870X.
52. European Standards—CEN. *EN1071-3/2005—Advanced Technical Ceramics—Methods of Test for Ceramic Coatings—Part 3: Determination of Adhesion and Other Mechanical Failure Modes by a Scratch Test*; European Standards—CEN: Brussels, Belgium, 2005.
53. Meneve, J.; Havermans, D.; von Stebut, J.; Jennett, N.; Banks, J.; Saunders, S.; Camino, D.; Teer, D.; Andersson, P.; Varjus, S. The Scratch Test: Atlas of Failure Modes. In *World Tribology Congress: Abstracts of Papers*; Institution of Mechanical Engineers: London, UK, 1997; p. 495.
54. CDC. Available online: https://www.cdc.gov/infectioncontrol/pdf/guidelines/disinfection-guidelines-H.pdf (accessed on 3 November 2020).
55. WHO. Available online: https://apps.who.int/phint/pdf/b/7.5.9.5.8-Methods-of-sterilization.pdf (accessed on 3 November 2020).
56. ISO. *ISO 22196:2011—Measurement of Antibacterial Activity on Plastics and Other Non-Porous Surfaces*; ISO—International Organization for Standardization: Geneva, Switzerland, 2011.
57. ISO. *ISO 10993-5:2009—Biological Evaluation of Medical Devices—Part 5: Tests for In Vitro Cytotoxicity*; ISO—International Organization for Standardization: Geneva, Switzerland, 2009.
58. Popa, A.C.; Marques, V.M.F.; Stan, G.E.; Husanu, M.A.; Galca, A.C.; Ghica, C.; Tulyaganov, D.U.; Lemos, A.F.; Ferreira, J.M.F. Nanomechanical characterization of bioglass films synthesized by magnetron sputtering. *Thin Solid Film.* **2014**, *553*, 166–172. [CrossRef]
59. Besleaga, C.; Dumitru, V.; Trinca, L.M.; Popa, A.C.; Negrila, C.C.; Kołodziejczyk, Ł.; Luculescu, C.R.; Ionescu, G.C.; Ripeanu, R.G.; Vladescu, A.; et al. Mechanical, corrosion and biological properties of room-temperature sputtered aluminum nitride films with dissimilar nanostructure. *Nanomaterials* **2017**, *7*, 394. [CrossRef]
60. Chemical Books Phosphorus Pentoxide. Available online: https://www.chemicalbook.com/ProductChemicalPropertiesCB9245033_EN.htm (accessed on 19 November 2020).
61. Wagner, C.D.; Riggs, W.M.; Davis, L.E.; Moulder, J.F.; Muilenberg, G.E. (Eds.) *Handbook of X-Ray Photoelectron Spectroscopy: A Reference Book of Standard Spectra for Identification and Interpretation of XPS Data*; Physical Electronics Division, Perkin-Elmer Corporation: Eden Prairie, MN, USA, 1992; ISBN 978-0-96-270262-4.
62. Carli, R.; Bianchi, C.L. XPS analysis of gallium oxides. *Appl. Surf. Sci.* **1992**, *74*, 99–102. [CrossRef]
63. Reed, T.B. *Free Energy of Formation of Binary Compounds: An Atlas of Charts for High-Temperature Chemical Calculations*; MIT Press: Cambridge, MA, USA, 1972; ISBN 978-0-26-218051-1.
64. Serra, J.; Gonzalez, P.; Liste, S.; Serra, C.; Chiussi, S.; Leon, B.; Perez-Amor, M.; Ylanen, H.O.; Hupa, M. FTIR and XPS studies of bioactive silica based glasses. *J. Non-Cryst. Solids* **2003**, *332*, 20–27. [CrossRef]
65. Nesbitt, H.W.; Bancroft, G.M.; Henderson, G.S.; Sawyer, R.; Secco, R.A. Direct and indirect evidence for free oxygen (O^{2-}) in MO-silicate glasses and melts (M = Mg, Ca, Pb). *Am. Mineral.* **2015**, *100*, 2566–2578. [CrossRef]
66. Ziemath, E.C.; Aegerter, M.A. Raman and Infrared investigations of glass and glass-ceramics with composition $2Na_2O{\cdot}1CaO{\cdot}3SiO_2$. *J. Mater. Res.* **1994**, *9*, 216–225. [CrossRef]
67. Agathopoulos, S.; Tulyaganov, D.U.; Ventura, J.M.G.; Kannan, S.; Karakassides, M.A.; Ferreira, J.M.F. Formation of hydroxyapatite onto glasses of the CaO–MgO–SiO_2 system with B_2O_3, Na_2O, CaF_2 and P_2O_5 additives. *Biomaterials* **2006**, *27*, 1832–1840. [CrossRef] [PubMed]
68. Xu, X.; Li, J.; Yao, L. A study of glass structure in Li_2OSiO_2, $Li_2OAl_2O_3SiO_2$ and LiAlSiON systems. *J. Non-Cryst. Solids* **1989**, *112*, 80–84. [CrossRef]
69. Zhang, Z.; Wang, W.; Korpacz, A.N.; Dufour, C.R.; Weiland, Z.J.; Lambert, C.R.; Timko, M.T. Binary Liquid mixture contact-angle measurements for precise estimation of surface free energy. *Langmuir* **2019**, *35*, 12317–12325. [CrossRef]

70. Lotfi, M.; Nejib, M.; Naceur, M. Cell Adhesion to Biomaterials: Concept of Biocompatibility. In *Advances in Biomaterials Science and Biomedical Applications*; Intech: Rijeka, Croatia, 2013; pp. 207–240. [CrossRef]
71. Popescu, A.C.; Florian, P.E.; Stan, G.E.; Popescu-Pelin, G.; Zgura, I.; Enculescu, M.; Oktar, F.N.; Trusca, R.; Sima, L.E.; Roseanu, A.; et al. Physical-chemical characterization and biological assessment of simple and lithium-doped biological-derived hydroxyapatite thin films for a new generation of metallic implants. *Appl. Surf. Sci.* **2018**, *439*, 724–735. [CrossRef]
72. Gentleman, M.M.; Gentleman, E. The role of surface free energy in osteoblast-biomaterial interactions. *Int. Mater. Rev.* **2014**, *59*, 417–429. [CrossRef]
73. Pešáková, V.; Kubies, D.; Hulejová, H.; Himmlová, L. The influence of implant surface properties on cell adhesion and proliferation. *J. Mater. Sci. Mater. Med.* **2007**, *18*, 465–473. [CrossRef]
74. Webb, K.; Hlady, V.; Tresco, P.A. Relative importance of surface wettability and charged functional groups on NIH 3T3 fibroblast attachment, spreading, and cytoskeletal organization. *J. Biomed. Mater. Res.* **1998**, *41*, 422–430. [CrossRef]
75. Ng, C.H.; Rao, N.; Law, W.C.; Xu, G.; Cheung, T.L.; Cheng, F.T.; Wang, X.; Man, H.C. Enhancing the cell proliferation performance of NiTi substrate by laser diffusion nitriding. *Surf. Coat. Technol.* **2017**, *309*, 59–66. [CrossRef]
76. Bellucci, D.; Bianchi, M.; Graziani, G.; Gambardella, A.; Berni, M.; Russo, A.; Cannillo, V. Pulsed electron deposition of nanostructured bioactive glass coatings for biomedical applications. *Ceram. Int.* **2017**, *43*, 15862–15867. [CrossRef]
77. KYOCERA. Available online: https://kyocera-sgstool.co.uk/titanium-resources/titanium-information-everything-you-need-to-know/titaniumproperties/#:~{}:text=It\T1\textquoterights%20Young\T1\textquoterights%20modulus%20of%20elasticity,shear%20modulus%20of%2045%20GPA (accessed on 3 November 2020).

Publisher's Note: MDPI stays neutral with regard to jurisdictional claims in published maps and institutional affiliations.

Article

The Surface Characterisation of Polyetheretherketone (PEEK) Modified via the Direct Sputter Deposition of Calcium Phosphate Thin Films

Shahzad Hussain, Leanne Rutledge, Jonathan G. Acheson, Brian J. Meenan and Adrian R. Boyd *

Nanotechnology and Integrated Bioengineering Centre (NIBEC), School of Engineering, University of Ulster, Shore Road, Newtownabbey, Co. Antrim, BT37 0QB. Northern Ireland, UK; s.hussain@ulster.ac.uk (S.H.); Rutledge-L1@email.ulster.ac.uk (L.R.); j.acheson@ulster.ac.uk (J.G.A.); bj.meenan@ulser.ac.uk (B.J.M.)

* Correspondence: ar.boyd@ulster.ac.uk; Tel.: +44-(0)2890-368924

Received: 16 October 2020; Accepted: 10 November 2020; Published: 13 November 2020

Abstract: Polyetheretherketone (PEEK) has emerged as the material of choice for spinal fusion devices, replacing conventional materials such as titanium and its alloys due to its ability to easily overcome a lot of the limitations of traditional metallic biomaterials. However, one of the major drawbacks of this material is that it is not osteoinductive, nor osteoconductive, preventing direct bone apposition. One way to overcome this is through the modification of the PEEK with bioactive calcium phosphate (CaP) materials, such as hydroxyapatite (HA–$Ca_{10}(PO_4)_6(OH)_2$). RF magnetron sputtering has been shown to be a particularly useful technique for the deposition of CaP coatings due to the ability of the technique to provide greater control of the coating's properties. The work undertaken here involved the deposition of HA directly onto PEEK via RF magnetron at a range of deposition times between 10–600 min to provide more bioactive surfaces. The surfaces produced have been extensively characterised using X-Ray Photoelectron Spectroscopy (XPS), Scanning Electron Microscopy (SEM), stylus profilometry, and Time of Flight Secondary Ion Mass Spectrometry (ToFSIMS). XPS results indicated that both Ca and P had successfully deposited onto the surface, albeit with low Ca/P ratios of around 0.85. ToFSIMS analysis indicated that Ca and P had been homogeneously deposited across all the surfaces. The SEM results showed that the CaP surfaces produced were a porous micro-/nano-structured lattice network and that the deposition rate influenced the pore area, pore diameter and number of pores. Depth profiling, using ToFSIMS, highlighted that Ca and P were embedded into the PEEK matrix up to a depth of around 1.21 µm and that the interface between the CaP surface and PEEK substrate was an intermixed layer. In summary, the results highlighted that RF magnetron sputtering can deliver homogenous CaP lattice-like surfaces onto PEEK in a direct, one-step process, without the need for any interlayers, and provides a basis for enhancing the potential bioactivity of PEEK.

Keywords: calcium phosphate; PEEK; surface modification; sputtering; ToFSIMS; XPS

---

1. Introduction

Polyetheretherketone (PEEK) is a high-performance thermoplastic polymer that has found increasing application in orthopaedic implant devices (such as spinal fusion cages), whereby its properties have been shown to outperform those of traditional metallic biomaterials, namely titanium alloys and stainless steel. PEEK is biocompatible, has enhanced resistance to in vivo degradation, favourable mechanical properties (such as stiffness, which is close to human bone, thereby reducing stress shielding), and it can be imaged without introducing artefacts (such as in X-Rays, Computed Tomography (CT) scanning and Magnetic Resonance Imaging (MRI)) [1]. Furthermore, PEEK has been shown to be a much easier material to work with than metals/metal alloys in terms of manufacturing, processability, cost and its ability to be easily 3D printed [2]. As such, it shows a lot of promise for orthopaedics (namely spinal fusion cages), and other 'made-to-measure' implants produced through additive manufacturing approaches [3]. A range of other approaches have also been considered for spinal repair, including studies by Gloria et al, whereby a polyetherimide (PEI)-based fusion cage reinforced with carbon fibres through filament winding and compression moulding provided structures with appropriate mechanical properties that would avoid stress shielding problems and other issues about metal ion release [4] In another study, poly(ε–caprolactone) intervertebral discs produced via additive manufacturing were shown to have favourable mechanical and in vitro properties [5]. Finally, work by Duarte et al developed 3D foams of polycaprolactone doped with polydopamine and polymethacrylic acid (PCL pDA pMAA) with appropriate mechanical properties which could be deployed without the use of instrumentation. [6] However, a key limitation of PEEK is that it is bioinert [7]. There is, therefore, a need to provide a mechanism to functionalise its surface, especially with respect to bone, to make the material at least osteoconductive to ensure a more rapid, improved, and stable fixation that will last longer in vivo. It has been considered that one way in which this can be achieved is through the modification of the PEEK implant with bioactive calcium phosphate (CaP) materials, such as hydroxyapatite (HA–$Ca_{10}(PO_4)_6(OH)_2$). HA is a highly valuable bone repair and regeneration material because of its similarity to the inorganic phase of human bone. It is bioactive, osteoconductive, and can form a direct chemical bond with human bone [8].

Several technologies have been utilised for the deposition of HA onto metals, including plasma spraying, electrophoretic deposition, pulsed laser deposition, sol-gel, biomimetic, and radio frequency (RF) magnetron sputtering [9]. Several methods have also been investigated as a means to deposit a bioactive HA coating onto PEEK, namely plasma spraying, [10] Ion Bean Assisted Deposition (IBAD) [11], aerosol deposition [12], spin coating [13], and RF magnetron sputtering [14]. Of these techniques, RF magnetron sputtering has shown significant promise for the deposition of CaP coatings due to the ability of the technique to provide greater control of the coating's properties and improved biological performance [15]. However, the previous studies, whereby HA was sputtered onto PEEK [16] (or IBAD onto PEEK) [17], required the use of an intermediate layer of yttria-stabilised zirconia (YSZ), thereby introducing additional processing steps, which adds complexity and enhanced cost to the method. Titanium dioxide and magnesium has also been sputter-deposited onto PEEK to alter the surface chemistry and morphology and influence their osteoblast cell behaviour and corrosion resistance, respectively [18]. There have also been studies whereby CaP materials have been sputtered onto polytetrafluorethane, polystyrene, polyethylene, polydimethylsiloxane, polylactic acid, and a copolymer of vinilidene fluoride and tetrafluoroethylene [19]. However, there have been no reports in the literature detailing the direct deposition of CaP materials onto PEEK using RF magnetron sputtering, to the knowledge of the authors.

This work was undertaken to study the RF magnetron sputter deposition of CaP materials onto PEEK, via a single step (direct) process, with the primary objective of creating a surface with specific chemistry and morphology commensurate with making the PEEK osteoconductive. Ideally, here the aim would be to deposit a HA coating, with properties commensurate with the requirements for HA coatings as laid out in the ISO (International Organisation for Standards) 13779-2 (2018) and ASTM (American Society for Testing and Materials (ASTM) F1609 standards. The work was completed using a custom designed RF magnetron sputtering facility utilising two sputtering targets (referred to as sources), operating at a low discharge power level (150 W). A low discharge power level was chosen for this study to prevent damage to the underlying polymer substrates and to ensure that the quality and consistency of the targets used could be guaranteed throughout the sputter deposition process. The effect of the deposition time on the surface morphology and chemistry were investigated here. All the surfaces produced were characterised using X-ray Photoelectron Spectroscopy (XPS), Scanning Electron Microscopy (SEM), optical profilometry, and Time of Flight Secondary Ion Mass Spectrometry (ToFSIMS). Therefore, this study represents the first attempt to deposit CaP materials directly onto PEEK using RF magnetron sputtering, and their subsequent surface characterisation.

2. Materials and Methods

2.1. Substrate Preparation

Circular Coupons of PEEK-OPTIMA™ LT1 (13 mm diameter and 2 mm thick) supplied by Invibio Ltd. (Thornton Cleveleys, UK) were abraded using 1200 P grade Silicon Carbide paper. Abrasion was carried out at approximately 250 revolutions per minute (RPM) for 3 min, removing all inhomogeneity from the substrate surface. The substrates were then twice sonicated in acetone (Sigma-Aldrich 99.5%, St. Louis, MI, USA) for 8 min and once in de-ionised water (DI) for 8 min and dried thoroughly in a convection oven at 70 °C for 12 h.

2.2. RF Magnetron Sputter Deposition

Sputtering targets were manufactured by dry pressing the hydroxyapatite (HA-(Plasma Biotal Captal-R), Tideswell, Buxton, UK) powder into low oxygen copper troughs (76 mm diameter and 5 mm thick) at a load of 40 kN for 10 min. RF Magnetron sputtering was undertaken using a designed Kurt J. Lesker Ltd., system (Hastings, UK), which was custom designed, operating with two Torus 3M sputtering sources operating at 13.56 MHz. The break-in prior to deposition from the HA target was conducted at a ramp rate of 5 watts (W) per minute up to the operating power of 150 W, whereby the source shutters were kept closed. The base pressure was below 5×10^{-6} Pa, with an argon gas flow rate (BOC, 99.995%) of between 15 and 20 Sccm, and a throw distance of 100 mm. During sputter deposition, the chamber pressure was maintained at 2 Pa. Table 1 outlines the sample nomenclature and key deposition parameters during the sputtering runs, with the deposition time being the main operating parameter varied during the experiments. The power density for these HA targets was approximately 3.3 $W{\cdot}cm^{-2}$ during each deposition.

Table 1. Sputter deposition operational parameters and sample nomenclature.

Sample Name	Deposition Time (min)	Pressure (Pa)	Power (W)
HA10	10	2	150
HA30	30	2	150
HA60	60	2	150
HA150	150	2	150
HA300	300	2	150
HA450	450	2	150
HA600	600	2	150

2.3. X-ray Photoelectron Spectroscopy

X-ray Photoelectron Spectroscopy (XPS) of the samples was undertaken out using a Kratos Axis Ultra DLD spectrometer (Manchester, UK). Spectra were recorded by employing monochromated Al Kα X-rays (hν = 1486.6 electron volts (eV)) operating at 15 kV and 10 mA (150 W). Wide energy survey scans (WESS) were obtained at a pass energy of 160 eV. High resolution spectra were recorded for O 1*s*, Ca 2*p*, P 2*p*, and C 1*s* at a pass energy of 20 eV. The Kratos charge neutraliser system was used on all samples with a filament current of 2.05 A and a charge balance of 3.8 V. Sample charging effects on the measured binding energy (BE) positions were corrected by setting the lowest BE component of the C 1*s* spectral envelope to 285.0 eV, [20–22]. Photoelectron spectra were further processed by subtracting a linear background and using the peak area for the most intense spectral line of each of the detected elemental species to determine the % atomic concentration. In total, 3 areas were analysed from each sample. Peak fitting was carried out using a mixed Gaussian–Lorentzian (GL (30)) synthetic peak function using the Kratos Vision software (version 2.3.0).

2.4. Time of Flight Secondary Ion Mass Spectrometry (ToFSIMS)

ToFSIMS data was obtained using a ToFSIMS IV instrument(ION-TOF GmbH, Münster, Germany) equipped with a 25 keV Bismuth (Bi) liquid metal ion gun (primary ion source) with a pulsed target current of 0.3 pico Amps (pA) and a post accelerator voltage of 10 kV, both with an incident angle of 45° to the sample surface normal. A base pressure of 6.66×10^{-6} Pa was maintained in the UHV analyser chamber during the analyses by the ToF method. The negative and positive secondary ion spectra were recorded with a Primary Ion Dose Density of 1×10^{13} ions/cm^{-2}. Charge compensation on the polymer surface was achieved using a low energy (21 eV) electron flood gun source, with the data acquired over a *m/z* range 0–200 for both positive and negative secondary ions. Data was presented by plotting *m/z* against intensity (counts/second). Ion images containing 256 × 256 pixels with 15 shots/pixel were acquired randomly, using Bi^{3+} primary ions in the high current bunched mode (HC-BU) over a 500 μm diameter area on the sample surface. Data acquisition and data processing and analysis were performed using SurfaceLab 6 (ION-TOF). The ToFSIMS IV instrument (ION-TOF GmbH, Münster, Germany) was also used to acquire depth profiles from the CaP modified PEEK samples, the instrument was equipped with a 20 keV Argon (Ar_{1900}^{+}) gas cluster ion gun which was rastered over the sample, with a crater size of 400 μm diameter, for 1000 secs. Analysis was completed using a Bi^{3+} liquid metal ion gun (primary ion beam) with an energy of 25 keV in a field of view (FOV) of 150 μm^2 and a raster size of 128 × 128 pixels (random mode). Due to the insulating nature of the samples, a low energy (21 eV) electron flood gun source was applied for the purposes of charge compensation with a filament current of 2.5 A. In total, 3 areas were analysed from each sample.

2.5. Optical Profilometry

A Zeta-20 Optical Profilometer (KLA Instruments, Milpitas, CA, USA) was used to undertake analysis of the craters that were produced as a result of sputtering during the ToFSIMS depth profiling experiments. The profile of each crater was measured using a 20× lens with a z range of 44 μm, 400 steps with a step size of 0.111 μm across 10-line profiles at a working distance of 3.1 mm. The FOV with the high resolution 0.35× coupler 2/3″ camera was 1169 × 876 μm^2.

2.6. Scanning Electron Microscopy Experimental Parameters

SEM analysis of the samples was carried out using a Dual Beam Quanta 200 3D (FEI, Hillsboro, UK). Prior to imaging the surface, it was coated (~12–17 nm) in gold-palladium to reduce surface charge effects. The sample surface was sputter coated with gold-palladium using an Emitech K500X SEM preparation sputter system. The images contained in this work were collected at an accelerating voltage of 10 kV and a working distance of 8–15 mm. The average pore area, number of pores, and Feret's diameter of the pores in each sputter deposited thin film were calculated using ImageJ software (Version 1.8.0) (National Institute of Health, Bethesda, MD, USA). In total, 3 areas were analysed from each sample.

2.7. Statistical Analysis

XPS, SEM, and ToFSIMS data in this study are reported as the mean ± standard deviation value (where $N = 3$). For the SEM analysis of the surfaces produced, a one-way analysis of variance (ANOVA) was applied to test for statistically significant differences between the sample types with a value of $P < 0.05$ considered to be statistically significant. The Bonferroni multiple comparison test was applied to compare values between successive pairs of sample types with the relevant outputs reported. All statistical analysis was performed using GraphPad Prism Version 3.0 software.

3. Results

3.1. SEM Analysis of Time Deposition Study

SEM analysis, as highlighted in Figure 1, was undertaken to examine the topographical nature of the pure PEEK substrates and the CaP thin film when the deposition time was varied. The SEM image for the pure PEEK substrate, highlighted in Figure 1a, illustrated a fairly flat surface with some scratches visible due to the abrasion of the surfaces. In comparison, the SEM images of HA modified PEEK (Figure 1b–h) indicated that there were marked differences between the topographies of each sample set (time deposition time points, 10–600 min, respectively). It is clear there were no voids in any of the thin films, and there was no evidence of cracking, delamination, and that the features shown below appear to be uniform. The initial interaction, as shown in Figure 1b (10 min deposition) and Figure 1c (30 min deposition), of the sputter deposited thin films and the PEEK material indicated that a physical change occurred on the surface of the PEEK material in the manner of 'pitting'. As the deposition time continued the SEM analysis indicated the development of a 'lattice-like' microstructure (Figure 1d–h). The average pore area (Figure 2a) and Feret's diameter (Figure 2b) for the pores was seen to generally increase with deposition time, until 450 min when they decreased. This was further corroborated with a calculation of the average number of pores per FOV as shown in Figure 2c. The number of pores declined with the increase in deposition time, which indicated that, with this coating process, there were fewer overall pores and that the pore area and diameter decreased, which indicated in-filling of the porous "lattice-like structure" as deposition time increased. The statistical comparison of the Feret's diameter, average pore area and number of pores is presented in Table 2, Table 3, and Table 4, respectively.

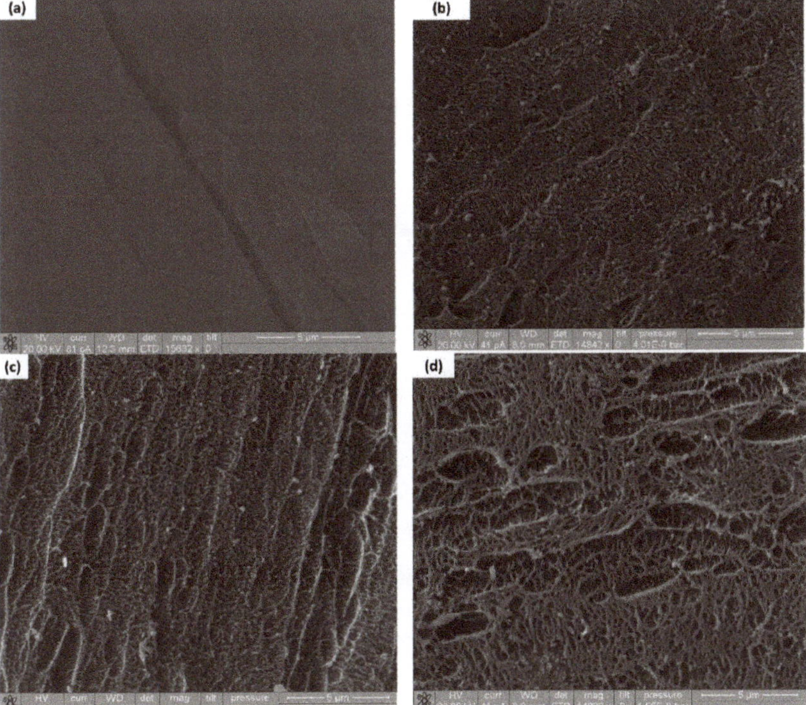

Figure 1. *Cont.*

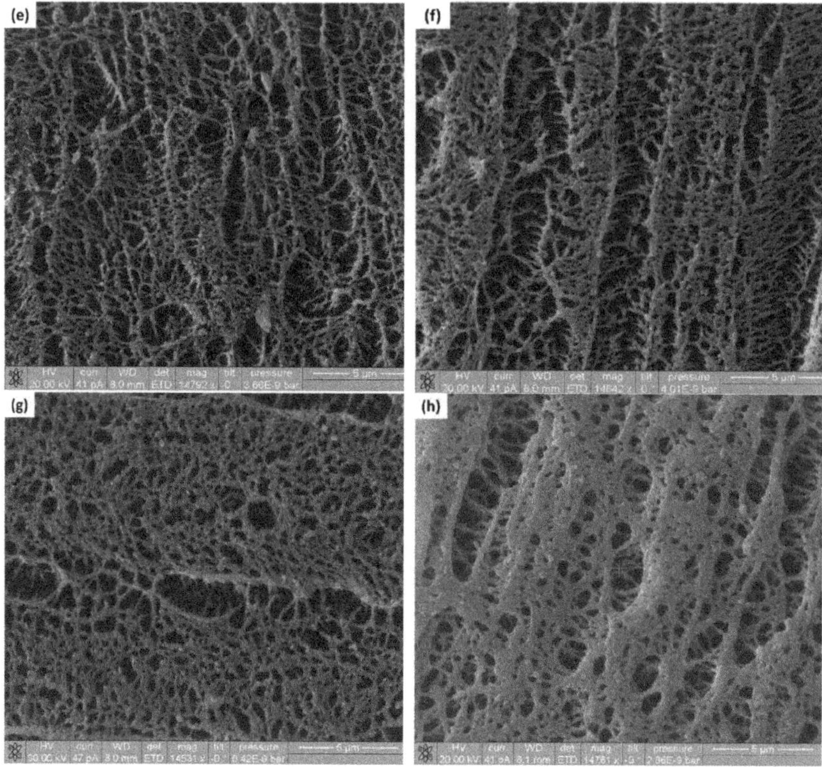

Figure 1. SEM images of, (**a**) PEEK substrate and the deposition of CaP onto PEEK for the samples (**b**) HA10, (**c**) HA30, (**d**) HA60, (**e**) HA150, (**f**) HA300, (**g**) HA450, and (**h**) HA600.

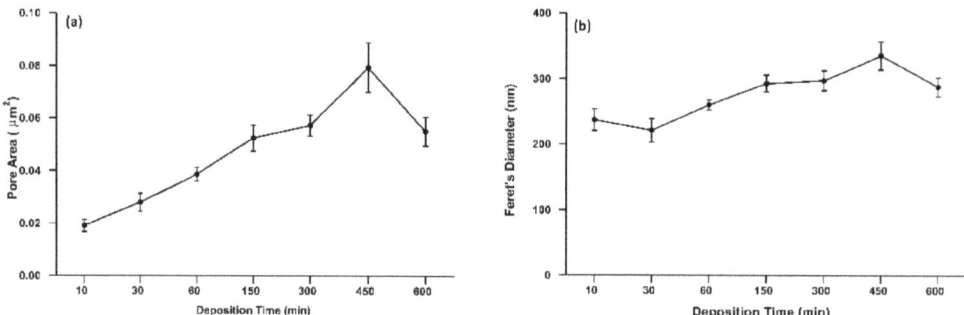

Figure 2. *Cont.*

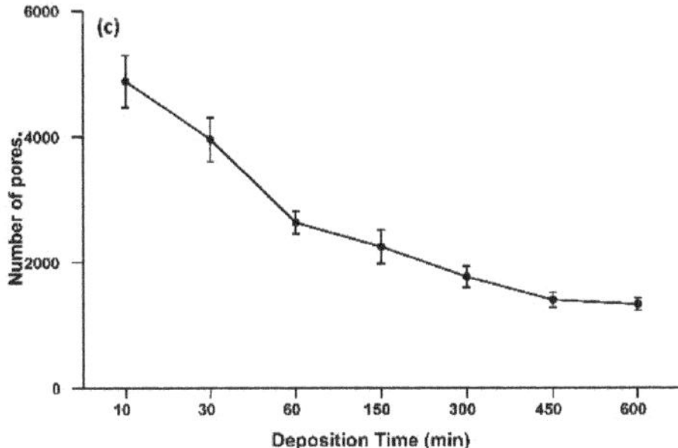

Figure 2. Results of the analysis of the SEM images showing (**a**). The average pore area of each modified PEEK sample; (**b**). The average Ferret's diameter of each modified PEEK sample; (**c**). The average number of pores for each modified PEEK samples.

Table 2. Statistical analysis of the average pore area of each across a variety of modified PEEK surfaces. $p > 0.05$: $p < 0.05$: *, $p < 0.01$: **, $p < 0.001$: ***.

Deposition Time (min)	Deposition Time (min)					
	30	**60**	**150**	**300**	**450**	**600**
10	**	***	***	***	***	***
30	-	***	***	***	***	***
60	-	-	***	***	***	***
150	-	-	-	-	***	-
300	-	-	-	-	***	-
450	-	-	-	-	-	***

Table 3. Statistical analysis of the average Feret's diameter of the pores across a variety of surfaces. $p > 0.05$: $p < 0.05$: *, $p < 0.01$: **, $p < 0.001$: ***.

Deposition Time (min)	Deposition Time (min)					
	30	**60**	**150**	**300**	**450**	**600**
10	-	*	***	***	***	***
30		***	***	***	***	***
60			***	***	***	**
150				-	***	-
300					***	-
450						***

Table 4. Statistical analysis of the number of pores across a variety of surfaces. $p > 0.05$: $p < 0.05$: *, $p < 0.01$: **, $p < 0.001$: ***.

Deposition Time (min)	Deposition Time (min)					
	30	**60**	**150**	**300**	**450**	**600**
10	***	***	***	***	***	***
30		***	***	***	***	***
60			*	***	***	**
150				**	***	***
300					*	**
450						-

3.2. XPS Analysis

The surface chemistry of the PEEK substrate, and CaP modified PEEK (HA10, HA60, HA300, and HA600) were all examined by XPS. The XPS analysis of the PEEK substrate, as shown in Figures 3 and 4 (the wide Energy Survey Scan (WESS) and high resolution scans of the C1s and O1s regions, respectively) reveals that the uppermost surface (<10 nm) consisted of C and O only. Table 5 shows the C and O concentrations (Atomic Concentration %) measured at 88.02 ± 0.37 and 11.98 ± 0.37 respectively, and an O/C ratio of 0.14 for the PEEK substrate. This compares favourably with an O/C ratio of 0.16/1 for PEEK as highlighted in the literature [2,23–25].

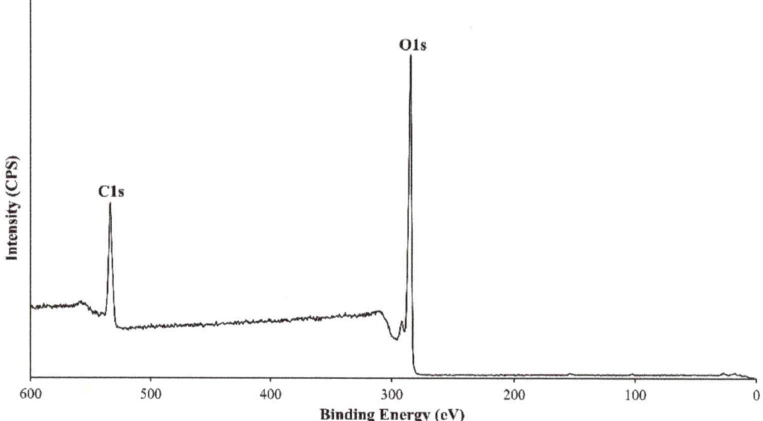

Figure 3. XPS survey scan (0–600 eV) for as-received PEEK.

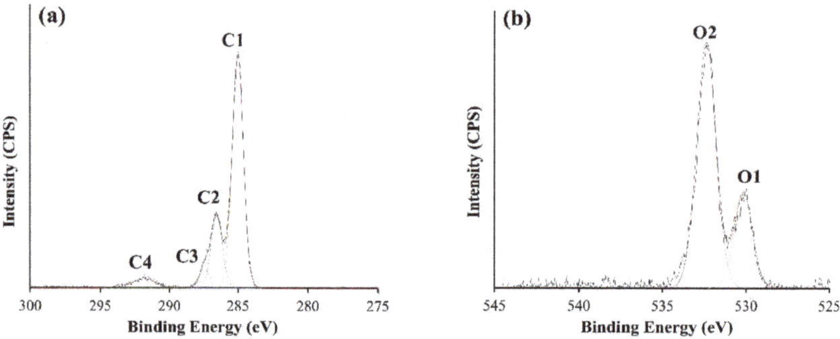

Figure 4. High-resolution peak fitted XPS spectra for as-received PEEK. (**a**) C 1*s* and (**b**) O 1*s*.

Peak fitted high-resolution C 1*s* and O 1*s* spectra of PEEK are illustrated in Figure 4. The O 1*s* envelope (Figure 4a) revealed two peaks, the most intense peak attributable to the ether, O–C group, at 533.2 eV (O2) (73%) whilst the peak located at 531.1 eV (O1) was found to be indicative of the carbonyl group, C=O, (27%) [25,26]. The C 1*s* core level spectrum, as shown in Figure 4b has been peak fitted using four peaks, of which the highest intensity peak has been noted at 285.0 eV (C1), C–H and C–C bonds [27–29]. A further peak was noted at 286.6 eV (C2) which is known to correspond to the binding of C and O atoms, C–O ether bonding. A shoulder at the higher B.E of 287.5 eV (C3) is indicative of aromatic C bound to O, O=C carbonyl bonds [24,25]. The theoretical composition of the different C bonding peaks were found in literature to be approximately 68.5% (C1), ××.7% (C2) and 5.8% (C3) [24,25,28], with the relative proportion of the three types of C bonding found within study to be very similar at 67% (C1), 21% (C2) and 6% (C3), as shown in Table 5. A low intensity peak at a 291.8 eV, 5% (C4), has been attributed to a shakeup satellite, occurring due to the presence of π–π* transitions [24,25,28]. No peaks representative of HA were observed on the PEEK substrate. Wide Energy Survey Scans and high-resolution XPS spectra of CaP as-deposited thin films onto PEEK, for (HA10) and (HA600) are presented in Figures 5–8.

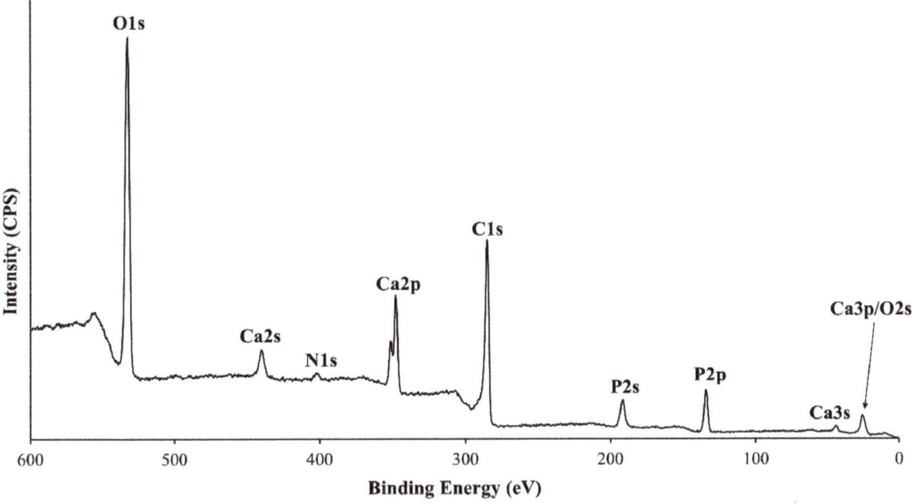

Figure 5. XPS survey scan (0–600 eV) for HA10.

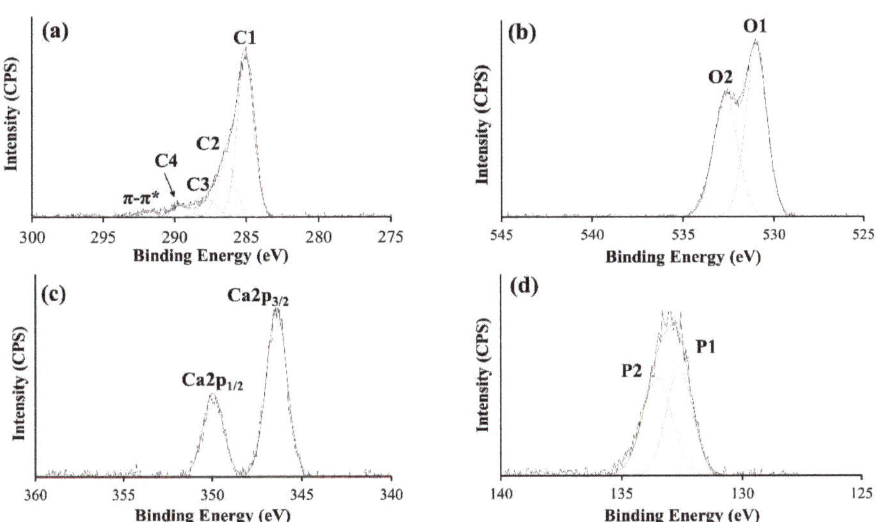

Figure 6. High-resolution peak fitted XPS spectra for the as-deposited HA10 thin film. (**a**) C 1s, (**b**) O 1s, (**c**) Ca 2*p*, and (**d**) P 2*p*.

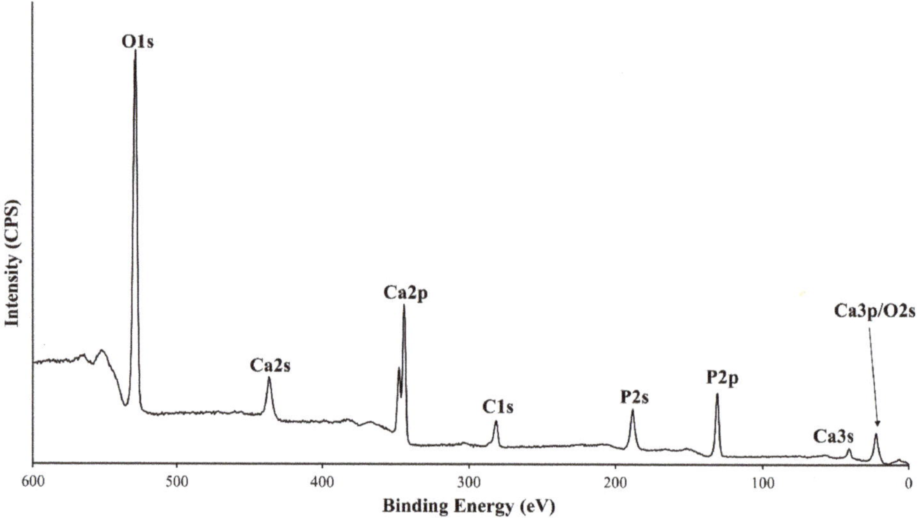

Figure 7. XPS survey scan (0–600 eV) for HA600.

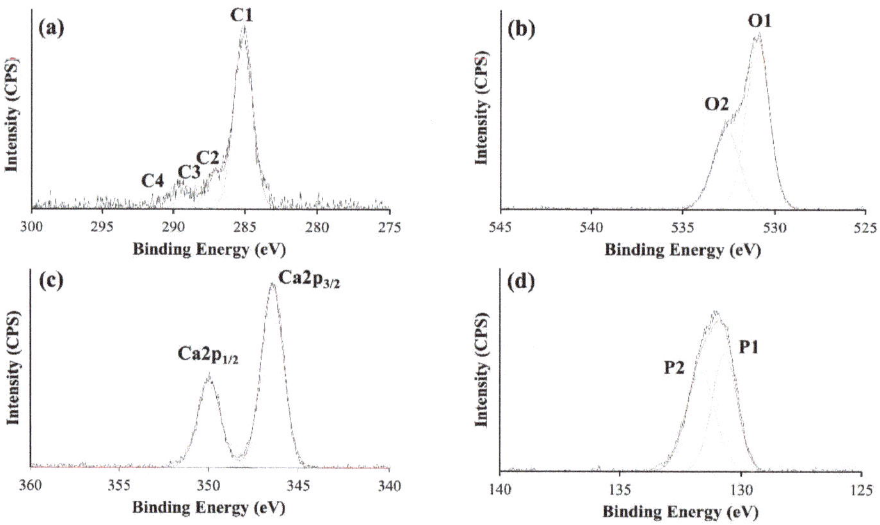

Figure 8. High-resolution peak fitted XPS spectra for the as-deposited HA600 thin film. (**a**) C 1*s*, (**b**) O 1*s*, (**c**) Ca 2*p*, and (**d**) P 2*p*.

Table 5 highlights the quantified result for the constituent elements of each of the high resolution O 1*s*, C 1*s*, Ca 2*p*, P 2*p* peaks including the Ca/P, and O/C ratios for each as-deposited surface. The C 1*s* envelope of the as-received PEEK HA10 thin film as shown in Figure 6a was observed as having three main peaks, representative of C–C/C–H, C–O, aromatic C=O, and CO_3^{2-} species at 285.0 (C1), 286.5 (C2), 287.8 (C3), and 289.5 eV (C4), respectively, with the sample having exhibited a peak at the higher B.E of 291.7 eV indicating the presence of a π–π* shakeup. It is worth noting that 300 (HA300) and 600-min (HA600) samples did not contain the π–π* shakeup peak. This can be observed in Figure 8a for the HA600 sample. The O 1*s* peak for the HA10 sample, illustrated in Figure 6b, exhibits two separate constituents, with the most intense observed at 531.0 eV (O1). This peak has been known to indicate O–P/O=P bonding and the less intense peak, at 532.7 eV (O2) has been attributed to OH^- species/O–C bonding. The high-resolution spectrum for a HA10 deposited thin film has revealed a Ca 2*p* envelope comprised of a well-resolved doublet with a B.E position for Ca $2p_{3/2}$ being observed at 346.5 and Ca $2p_{1/2}$ at 350.0 eV, as highlighted in Figure 6c. The fitted P 2*p* spectral envelope, as illustrated in Figure 6d, displayed B.E positions of 132.6 and 133.5 eV, which represents P–O (P1) and P–OH (P2) bonds, respectively.

The peak positions for the HA60, HA300, and HA600 samples were very similar to those of the HA10 thin film onto PEEK. As the deposition time increased, so too did the % AC of O 1*s*, Ca 2*p* and P 2*p*, while the concentration of each of the fine components of C 1*s* decreased, as can be seen in Table 5. The Ca/P ratio for each of the as deposited thin films were very similar being 0.83 ± 0.04, 0.84 ± 0.03, 0.85 ± 0.03, and 0.86 ± 0.04 for the 10, 60, 300, and 600 min surfaces, respectively. The O/C ratios for each of the as deposited thin films were 0.69 ± 0.02, 1.23 ± 0.05, 2.06 ± 0.08, and 4.00 ± 0.08 for each of the 10, 60, 300, and 600 min surfaces, respectively.

Table 5. Percentage of different of different elements and their ratios.

Elements	**Samples**			
	HA10	**HA60**	**HA300**	**HA600**
O 1*s* (%)	36.01 ± 0.48	44.91 ± 0.75	53.12 ± 0.78	58.46 ± 0.07
C 1*s* (%)	53.43 ± 0.53	37.85 ± 1.12	25.83 ± 0.59	14.63 ± 0.28
Ca 2*p* (%)	4.78 ± 0.11	7.87 ± 0.32	9.50 ± 0.29	12.47 ± 0.23
P 2*p* (%)	5.77 ± 0.25	9.38 ± 0.20	11.54 ± 0.16	14.43 ± 0.37
Ca/P ratio	0.83 ± 0.04	0.84 ± 0.03	0.85 ± 0.03	0.86 ± 0.04
O/C ratio	0.69 ± 0.02	1.23 ± 0.06	2.06 ± 0.08	4.0 ± 0.08

3.3. ToFSIMS Study

In order to determine the presence or otherwise of particular chemical species, ToFSIMS was employed to provide information about the chemical nature of the outermost molecular layer (1–10 Å) of the PEEK and CaP modified PEEK (HA10, HA300 and HA600) samples via a mass survey. Given the high mass resolution of SIMS it was particularly well suited to polymer surface analysis; however, the analyses of PEEK via ToFSIMS has been scarcely described within the literature [27,28]. The positive and negative ion spectra for the PEEK polymer have been acquired in the relevant *m/z* ranges of 0–200. For the purposes of highlighting key results, the positive and negative ion spectra have been shown here between *m/z* ranges of 25–110 and 20–100, respectively.

3.3.1. ToFSIMS Analysis of PEEK

From the positive survey for the PEEK substrate, as shown in Figure 9a, it can be noted that peaks above an *m/z* of 100 were very weakly detected. Specific peaks with a high intensity included *m/z* 39, 51, 77, 91, 104, 105, 115, 139, 152, 163, 165, and 195–197, these were all considered to be either indicative of aromaticity or ionically diagnostic of PEEK/PEEK fragments by Pawson et al. [28]. The peak at *m/z* 165 has been noted as useful in determining polymer structure as it contains no O [30]. In comparison, for the negative ion spectrum for the PEEK substrate (shown in Figure 9b), a range of peaks indicative of PEEK fragments were detected, including *m/z* 25, 41, 49, 73, 108, 121, and 196, which were recorded as corresponding to C_2H^-, C_2OH^-, C_4H^-, C_6H^-, $C_6H_4O_2{}^-$, $C_7H_5O_2{}^-$, and $C_{13}H_8O_2{}^-$, respectively. The peak at the *m/z* ratio of 197 was assigned to the repeating structure of PEEK polymer.

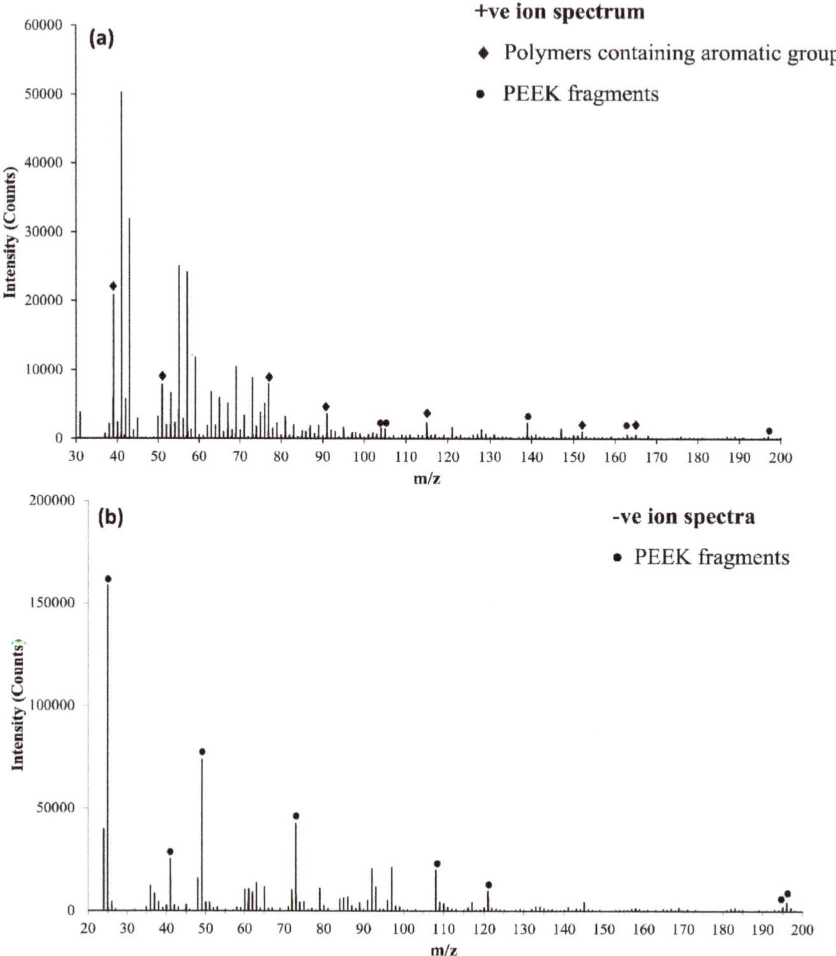

Figure 9. ToFSIMS survey spectra for (**a**) positive ions for the PEEK substrate, (**b**) negative ions for the PEEK substrate.

Positive ion intensity surface maps for the PEEK substrate surfaces have been recorded in Figure 10a. The peaks at *m/z* 40 and 57 in Figure 10a (i) and (ii) have been known to represent Ca^+ and $CaOH^+$, respectively. Low levels of both ions and fragments, in comparison to a modified surface have been exhibited. The surface maps recorded for *m/z* 104, 139, and 163 (Figure 10a (iii)–(iv), respectively) are thought to be indicative and diagnostic of PEEK. Figure 10b has presented the negative ion intensity surface maps for the PEEK surfaces at the *m/z* detailed. The peaks at *m/z* 63 and 79 have been known to represent PO_2^- and PO_3^-, (Figure 10b (i) and (ii)), respectively, low levels of both have been found to be present in comparison to a CaP modified surface. Due to handling and the nature of the sampling technique, some surface contamination was not unexpected. *M/z* 108, 121, and 196 peaks (Figure 10b (iii)–(iv)) have been assigned to the $C_6H_4O_2^-$, $C_7H_5O_2^-$, and $C_{13}H_8O_2^-$ fragments, respectively; these were found to be present in a higher intensity in comparison to the modified surface. The images have indicated homogeneity of the selected ions (150 µm FOV) on the surface of the PEEK material. No unexpected species were detected; the samples were found to be relatively free from contamination.

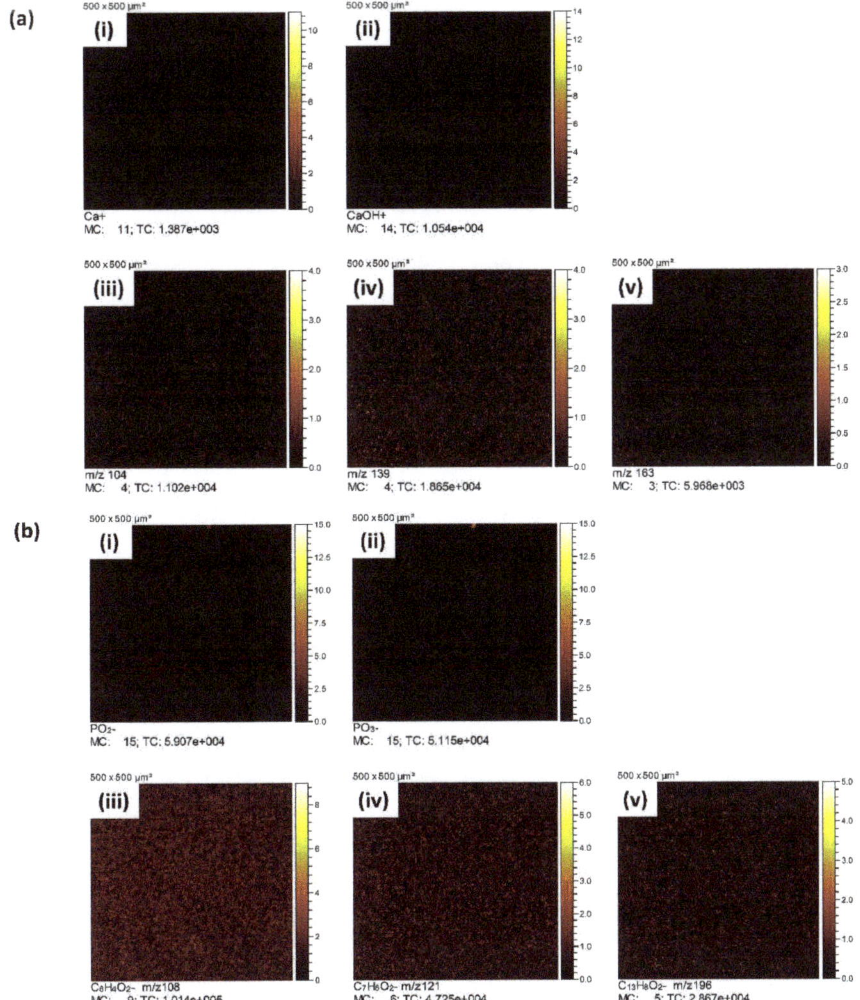

Figure 10. (**a**). ToFSIMS positive ion intensity surface maps for PEEK obtained at selected masses (i) Ca^+, (ii) $CaOH^+$, (iii) *m/z* 104, (iv) *m/z* 139 and (v) *m/z* 163. All images were normalised to the total ion count; (**b**) ToFSIMS negative ion intensity surface maps for PEEK obtained at selected masses (i) PO_2^-, (ii) PO_3^-, (iii) $C_6H_4O_2^-$, 108 (iv) $C_7H_5O_2^-$ 121, and (v) $C_{13}H_8O_2^-$ 196). All images were normalised to the total ion count.

3.3.2. ToFSIMS Analysis of CaP Deposited onto PEEK over Time

The positive ion survey spectra of the CaP thin film sputter deposited onto the PEEK surface for up to 10 min (HA10), in the *m/z* 25–110 range is shown in Figure 11a, with characteristic peaks observed at an *m/z* ratio of 40 and 57 representative of Ca^+ and $CaOH^+$, respectively [9]. Isotopes of Ca^+ were detected at *m/z* 42 and 44, and it was clear from looking at the survey spectra that the $CaOH^+$ ion was dominant in comparison to the neat PEEK spectra. The presence of impurity ion K^+ was noted at *m/z* 39. Peaks known to relate to polymers containing aromatic groups were found to be present at *m/z* 39, 51, 77, and 91. At *m/z* ratios above 100, there were two peaks, namely at 104 and 105, which were thought to be specific to PEEK [28] in this circumstance, and only the peak at 105 was considered to be significant

(>0.4% of the largest peak) [9]. The negative ion survey of the CaP thin films sputter deposited onto the PEEK surface for up to 10 min in the *m/z* 20–100 range is shown in Figure 11b, with characteristic peaks within the spectra noted at an *m/z* 63 and 79 corresponding to PO_2^- and PO_3^-. Peaks were also observed at 25, 41, 49, and 73, corresponding to the main negative fragments of PEEK, C_2H^-, C_2OH^-, C_4H^- and C_6H^- [27], these have been attributed to either PEEK or surface contamination due to their exposure to atmospheric conditions. Similar results were observed for figure the positive and negative ion spectra for the HA600 sample, as shown in Figure 12a,b, respectively. It is noted that the relative intensities of the PO_2^- and PO_3^- ions dominant the negative ion spectra for the HA600 surface, with the contribution from the PEEK significantly diminished. The ToFSIMS surface mapping positive and negative analysis for HA600 has been presented in Figure 13a,b, respectively, with similar results observed when compared to those of the HA10 samples in Figure 10. The positive and negative peak area (normalised by total ion count) bar charts for of the samples analysed here by ToFSIMS are shown in Figure 14a,b.

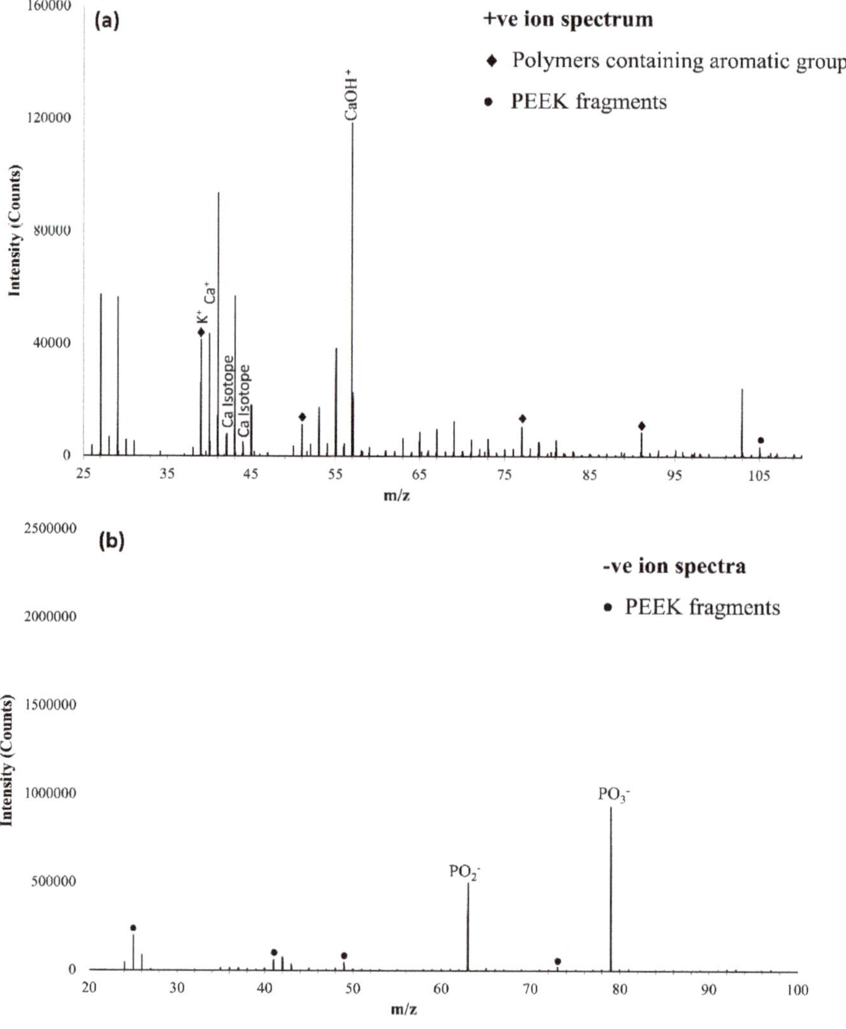

Figure 11. ToFSIMS survey spectra for (**a**) positive ions HA10, (**b**) negative ions HA10.

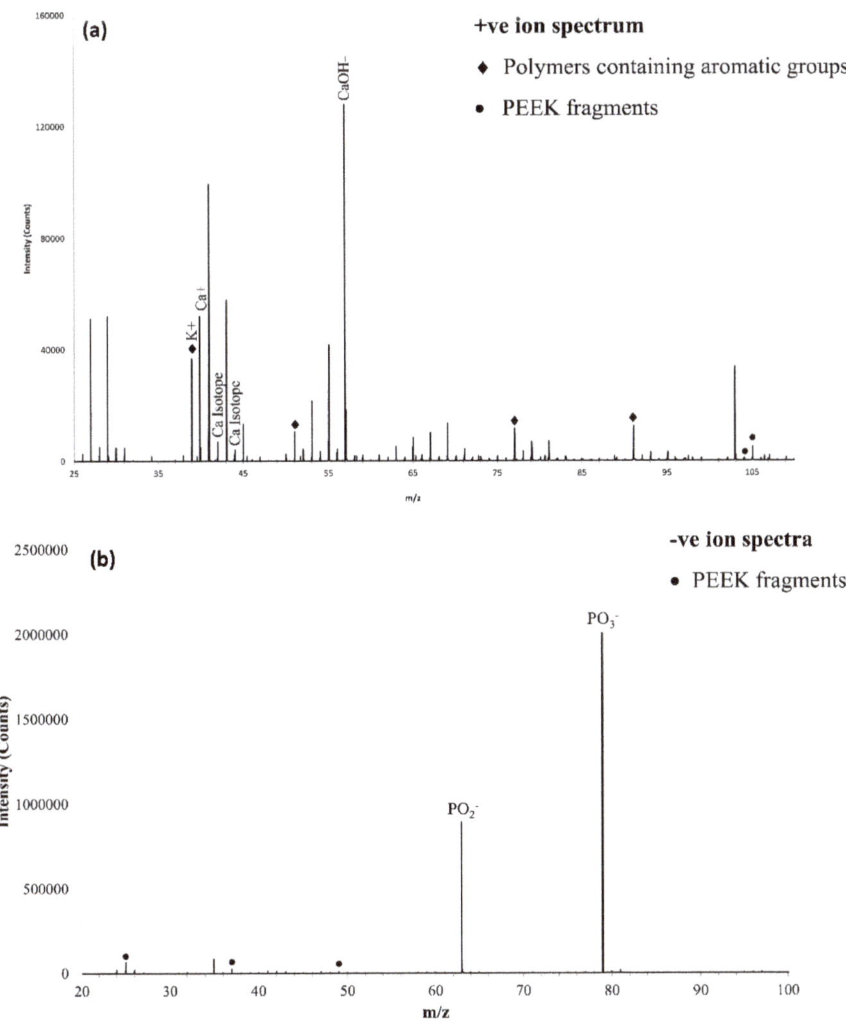

Figure 12. ToFSIMS spectra for (**a**) positive ions HA600 and (**b**) negative ions HA600.

3.3.3. ToFSIMS Depth Profile of CaP Sputter Coated PEEK (HA600)

To further investigate the relationship between the sputtered surface and the PEEK substrate, ToFSIMS depth profiling was employed. The depth profile for HA600 modified PEEK in the positive and negative ion modes (1000 s rastering) is shown in Figure 15 (overlaid for both positive and negative ions). There was a decline in intensity of the P and Ca ions as they appeared to tail off in response to the Ar^+ ion bombardment. There was a very slight, yet steady and continuous increase in the intensities correlating to the PEEK fragments, both negative and positive. These ions were expected to be in low intensity, whereas the hydrocarbon was expected to be much more intense. The ion intensity of the C_2H^- hydrocarbon experienced an initial sharp decline at the very beginning of rastering; this was likely due to surface contamination, and was followed by an incline, followed by an exponential decrease to a less intense but steadier state. A further depth profile, taken using ToFSIMS, is shown in Figure 16a, probing only the PO_3^- and C_4H^- ions (as shown in Figure 16b,c) representative of the CaP coating and PEEK substrate, respectively), highlights the fact that the there is significant intermixing of the coating and the PEEK substrate. The depth profilometry of the ToFSIMS sputter crater is illustrated in Figure 17 It was found that the sputter crater was, on average, 2.32 ± 0.19 µm.

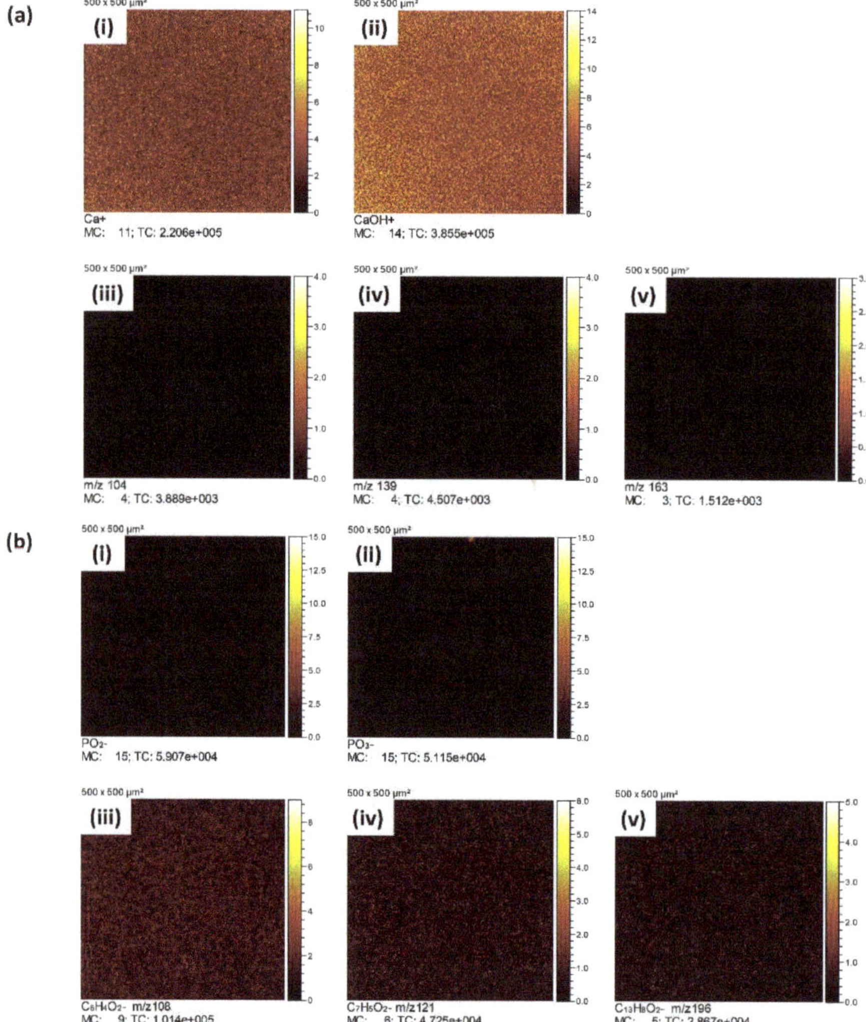

Figure 13. (**a**) ToFSIMS positive ion intensity surface maps for HA600 obtained at selected masses (i) Ca^+, (ii) $CaOH^+$, (iii) *m/z* 104, (iv) *m/z* 139 and (v) *m/z* 163. All images were normalised to the total ion count; (**b**) ToFSIMS negative ion intensity surface maps for HA600 obtained at selected masses (i) PO_2^-, (ii) PO_3^-, (iii) $C_6H_4O_2^-$, 108 (iv) $C_7H_5O_2^-$ 121, and (v) $C_{13}H_8O_2^-$ 196). All images were normalised to the total ion count.

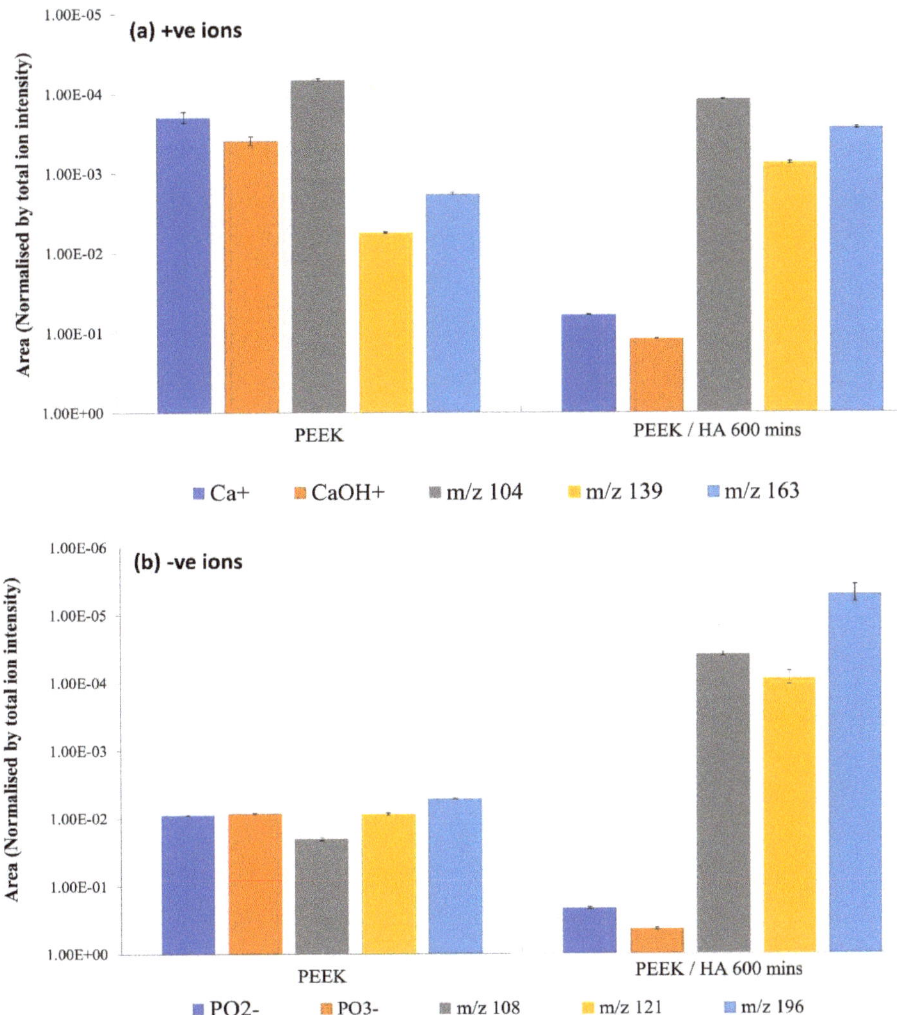

Figure 14. (**a**) PEEK substrate and HA600 positive ion intensity for the Ca^+, and $CaOH^+$, *m/z* 104, 139, and 163 (diagnostic PEEK fragments); (**b**) PEEK substrate and HA600 negative ion intensity for the PO_2^- and PO_3^- ions, $C_6H_4O_2^-$ (*m/z* 108), $C_7H_5O_2^-$ (*m/z* 121), and $C_{13}H_8O_2^-$ (*m/z* 196).

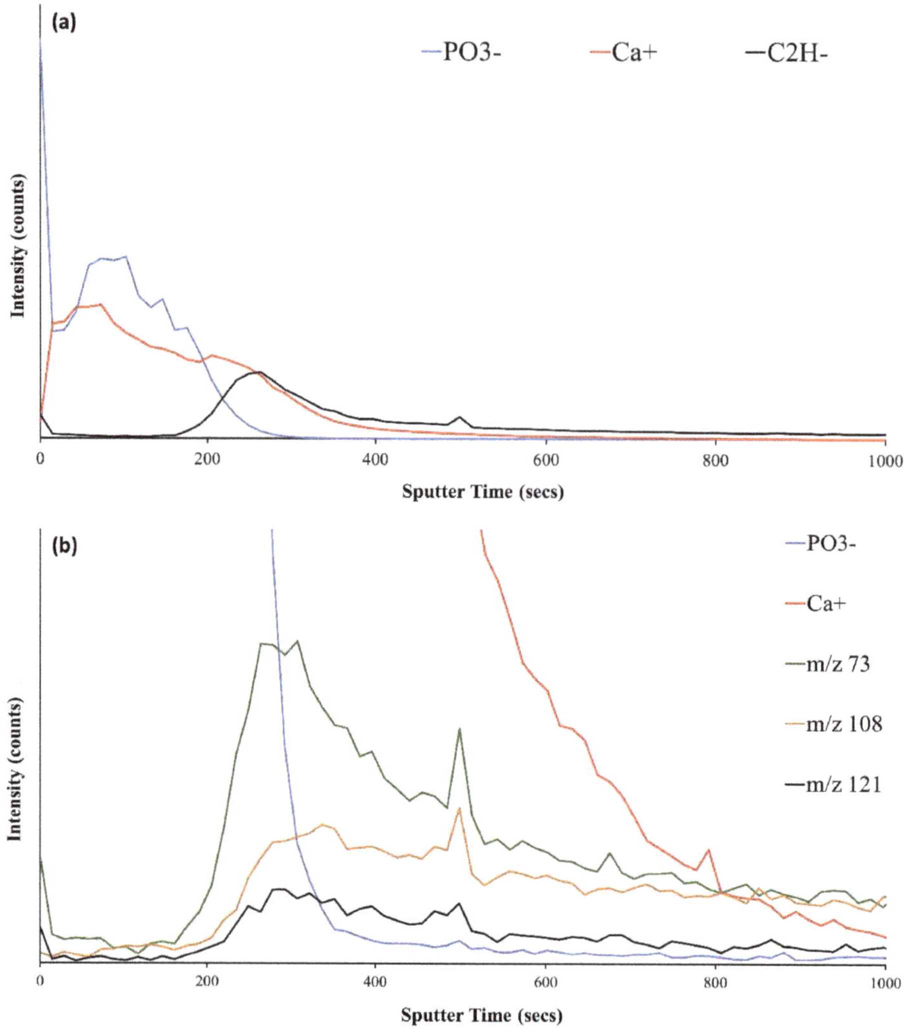

Figure 15. ToFSIMS depth profiles of CaP-modified PEEK (HA600). Positive and negative spectra are overlaid, {(**a**) 100,000 counts, (**b**) 1000 counts.}.

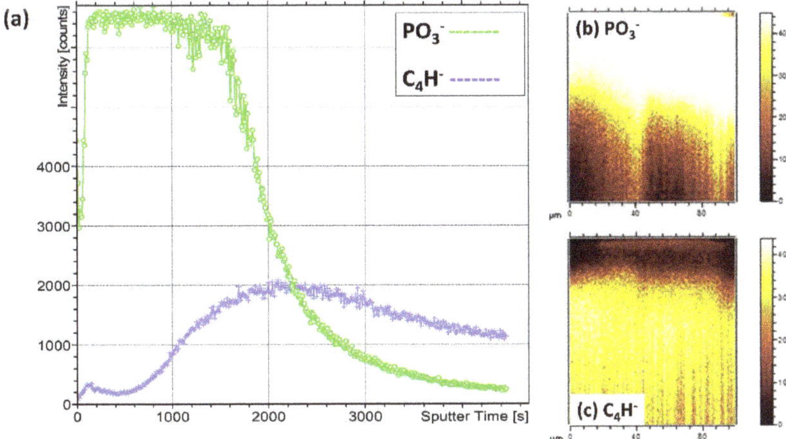

Figure 16. (**a**) ToFSIMS depth profile for PO_3^- and C_4H^- ions, with 2D images for (**b**) the PO_3^- and (**c**) the C_4H^- ions.

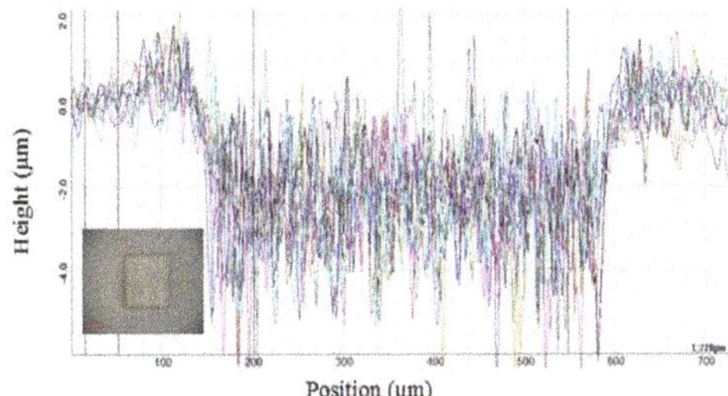

Figure 17. Optical Profilometry cross section of the HA600 depth profile.

4. Discussion

RFMS was successfully utilised to modify the surfaces of PEEK polymer at low temperatures with a CaP thin film. This was done with the objective of potentially enhancing the PEEK material's surface bioactivity in preparation for exposure to physiological conditions. Through the manipulation of the sputtering parameters, namely deposition time, this work has shown that the desired CaP thin films can be achieved successfully, in a one step process, without the need for subsequent processing (e.g. thermal annealing), albeit at a CaP ratio below that expected for pure HA. To ascertain the potential of these surfaces, several different analytical techniques were used to characterise the resultant thin films chemically and physically, including ToFSIMS, XPS, and SEM. The XPS data obtained for PEEK was typical of this material. Each peak assignment for the curve fitted elements were as expected [24–26,28,29,31]. The O/C value obtained for PEEK, 14/1, was lower than that which was anticipated (16/1) [2,23,25], small deviations such as this have been attributed to the presence of hydrocarbon contamination on the PEEK surface [28]. With regard to the ToFSIMS analysis of PEEK, according to Pawson et al. [28] and Lub et al. [30], specific peaks, within the survey, at *m/z* of 39, 51, 77, 91, 115, 152, and 165 have been considered as representative of aromaticity, such as the structure of PEEK. The peak at *m/z* 165 was originally found in a polycarbonate, which is known as structurally very similar to that of PEEK, this particular peak has been recorded as being useful in determining the polymer structure as it contained no O, due to the polymer being analysed after rearrangement [30]. Peaks outlined by Pawson et al. [28] which were thought to be indicative of PEEK included *m/z* 104, 105, 139, 163, and 195–197. The peak at an *m/z* ratio of 39 has been treated with care as, according to literature, some of the intensity has been known to be attributed to potassium as a result of remnants left from the polymerisation process. It was not detected via XPS here. According to Pawson et al. [28], peaks below *m/z* 60 within the negative survey have not been found to be useful diagnostically and intensities are low, which, from looking at other literature, was expected. Work by Henneuse-Boxus et al. [27] has detailed a number of peaks for the *m/z* ratios of PEEK fragments which have been found within the negative survey presented here, these included 25, 41, 49, 73, 108, 121, and 196, which were recorded as corresponding to C_2H^-, C_2OH^-, C_4H^-, C_6H^-, $C_6H_4O_2^-$, $C_7H_5O_2^-$ and $C_{13}H_8O_2^-$, respectively. The peak at the *m/z* ratio of 197 has been assigned to the PEEK monomer. Positive ion intensity surface maps for the PEEK surfaces have included peaks present at *m/z* 40 and 57, which represent Ca^+ and $CaOH^+$, respectively. Low levels of both ion fragments, in comparison to a modified surface, were noted as expected due to this surface being essentially pure PEEK (PEEK Optima LT1 grade). This intensity was likely contributed to by a fragment other than Ca^+ or $CaOH^+$. Fragments present at *m/z* 104, 139 and 163 are known to be indicative and diagnostic of PEEK, these were present in higher intensities than those of the CaP modified surface, this was expected. These ions have been commonly found in low intensities [28]. The ToFSIMS negative ion intensity surface maps for PEEK surfaces at the selected masses exhibited low levels of PO_3^- and PO_2^-, in comparison to a CaP-modified surface, as expected due to the fact that this particular surface has had no exposure to any modification procedure. The surface maps for *m/z* 108, 121, and 196 have been associated with aromatic polymers or fragments of PEEK [27], it was expected that these would be present in a higher intensity than they were found, in comparison to a modified surface. C_nH_m clusters were observed in much higher intensities than the peaks chosen (*m/z* 108, 121, and 196) and are known as attributable to carbonaceous species, these have been thought of as generally indicative of a polymeric material; however, these were considered too generic to utilise on this occasion as, according to literature, they have not been considered diagnostic of PEEK [28]. SEM micrographs of PEEK modified with CaP via RFMS for various deposition times have been shown in Figure 1. The morphology of the surface of the polymer was characterised for time periods up to 600 min, during which time the polymer surface morphology developed, from the early time point of 10 min pitting of the surface could be clearly observed, whereas a more intricate lattice-like or network-type microstructure was exhibited at later time points. It was thought that this development was a consequence of the eventual coalescence of the pits to establish porosity via the nucleation and growth of the surface structures. The size of the pits, according to the area and Feret's diameter calculations, has been shown to rise as time increased, until 450 min, when the dimensions started to decrease. This, along with the knowledge that the number of pores declined as time deposition increased, indicated that an infilling of the surface morphology has occurred after 450 min. The nature of the porous microstructure exhibited at latter deposition times was similar in nature to previous work [32–35] on polymers. As such, the surface of the PEEK substrates obviously undergo etching at the start of the sputtering process, with Ca and P species then becoming embedded into the PEEK and the resultant CaP coating nucleating from these embedded species, resulting in a hybrid intermixed zone between the coating and the substrate. Evidence for this hybrid intermixed CaP coating and PEEK substrate zone is shown by the cross-sectional depth profile illustrated in Figure 16, where there is clear overlap between the PO_3^- and C_4H^- ions species (representative of the CaP coating and the PEEK substrate, respectively). It is suggested that, as the Ca and P species do not embed homogeneously over the entire surface during sputtering, this results in the formation of the lattice-like network as opposed to a normal dense, non-porous and continuous coating as normally observed with such coatings on metallic substrates [20].

In order to investigate the nature of the CaP thin films once they had been deposited onto the PEEK substrate, for depositions times 10, 60, 300, and 600 min, ToFSIMS was employed including the survey, surface mapping and depth profiling functionalities. XPS was used to investigate the chemistry at the time points (10, 60, 300, and 600 min). The positive ToFSIMS surveys were displayed in Figures 11a and 12a for the samples HA10 and HA600, respectively. Each of the spectra displayed the characteristic peaks expected for a ToFSIMS analysis of a CaP modified surface [9,19,36]. In each spectrum, Ca^+ species (*m/z* 40 and 57) as well as the isotopes (*m/z* 42 and 44) were present. Various peaks within the positive spectrum considered to be representative of polymers with aromatic groups were present (*m/z* 39, 51, 77, and 91) along with peaks thought indicative of PEEK (*m/z* 104 and 105). The negative ToFSIMS spectra for both the HA10 AND HA600 samples (Figures 11b and 12b respectively), revealed the dominance of the key ion, P (*m/z* 63 and 79) for both time points. ToFSIMS results revealed that, as the length of time that the CaP material was sputter deposited onto the surface of the PEEK substrate increased, the relative intensity of the PEEK fragments diminished in both the positive and negative spectra, whilst the relative Ca and P intensities increased as shown in Figure 14. These results were consistent with the XPS results. The WESS and high-resolution spectra for XPS for the HA10 and HA600 samples (shown in Figures 5–8) exhibited peaks which were characteristic of CaP materials and were as expected [9,12,37]. It was noted from these results that the Ca/P was very low for all the samples and well below the expected 1.67. Previous work by Surmenev et al., showed that the Ca/P ratio was much higher when HA was sputter deposited onto polymeric materials (PTFE), and that it was difficult to form CaP coatings on such substrates [38]. The results here are contradictory to those previous findings and are most likely due the build-up of negative charge on the surface of the insulating PEEK substrate. This prevents re-sputtering of P by negatively charged O, resulting in the low Ca/P observed here [39]. ToFSIMS surface mapping for both the PO_2^- and PO_3^- and Ca ions has shown a similar trend. Incidentally, it was also reported by various authors that both Ca and P travel from the target to the substrate as neutral species [38] depositing to form CaP thin films [40–43]. When analysing the ToFSIMS spectra, it was noted that PO_2^- and PO_3^- ions were dominant ions in the negative ToFSIMS survey, and with increasing deposition time their presence increased, correlating well with the increase in both O and P highlighted in the corresponding XPS. This information, along with the decrease in the C % AC, with time, caused the O/C ratio to increase. High levels of C present on the CaP–modified PEEK surfaces have been reported in literature [9]. It was considered that this level of C may have been due to 'adventitious C contamination' because of the adsorption of impurity hydrocarbons, which corroborated the ToFSIMS results. ToFSIMS analysis indicated that, in both the positive and negative surveys, the peaks thought to be representative of hydrocarbon clusters (C_nH_m) had high intensities, (*m/z* 29, 49, and 73); however, as deposition time increased, the intensity of these peaks depreciated. It was also considered that some of the intensity of the C 1*s* high resolution XPS scan may have been attributed to C–C/C–H bonds. It was postulated from analysing XPS and ToFSIMS data that the decrease in C intensity was attributable to the sputter deposited thin film growing with time and, therefore, fewer PEEK C–C/C–H bonds were being detected, as the detection depth limits of the XPS system was limited to ~5–10 nm. After 600 min of sputter deposition it was realised that a number of the peaks defined within literature as being specific and diagnostic of PEEK, *m/z* 104 and 105, were still being detected by ToFSIMS (detection limit of 1–2 nm); however, only the 105 peak was considered significant (>0.4% of largest peak) [9]. It was suggested that this peak may have been due to the presence of another organic species, as corroborated by XPS, and the porous nature of the surfaces produced; however, this does merit further investigation. The shakeup peak often affiliated with aromatic ring structures present in PEEK have been exhibited in the C1s high resolution survey scans for 10 and 60 min but have disappeared for the 300 and 600 min modifications. The XPS no longer detected the underlying polymer substrate from 300 min, which suggested that the thin film deposition had grown so that the polymer was out of the detection limits of the XPS instrument. ToFSIMS surface mapping showed a similar trend, whereby all six of the fragments for PEEK, in the positive and negative spectra, were reduced in comparison to the neat PEEK sample, while the Ca, PO_2^- and PO_3^- ions had increased intensities. The ratio of each bond within the C 1*s* scan to the total % AC of C was similar to that expected for PEEK for the 10 and 60 min deposition sample with the addition of the CO_3^{2-} bonds having possibly stemmed from the modification of the substrate with amorphous HA as outlined by Surmenev et al. [44]. There was strong evidence that suggested that there was a formation of a hybrid interaction layer between the PEEK and CaP material. In order to investigate the chemical nature of the interaction, ToFSIMS depth analysis was carried out, revealing that the P-related ions are readily detected on the surface of the modified PEEK substrate to a depth of

~ 0.65 μm, whereas the Ca ion signal was detected for much longer and appeared to tail off. This data indicated that sputtered species were possibly embedded into the polymer material up to a depth of ~1.21 μm, as determined by Zeta analysis. It was suggested that, once the Ca embedded into the polymer, P proceeds to grow on this Ca rich layer, with all species known to be eligible for re-sputtering in the dynamic RFMS environment. Within the known literature [30,45] several potential hypotheses are provided as to the reason for intensity spikes of each ion of interest. It was thought that, in the case of this work, the most likely cause was matrix ionisation effects that have been associated with surface pollutants, such as an oxidation layer. The evidence to substantiate this event comes from the fact that oxidation was thought to coincide with the small spike in intensity for signals at the beginning of each depth profile [46]. Another potential hypothesis that has been recorded was that the full depth profile regime has been met after a few seconds of sputtering time. The interface between the CaP film and the PEEK substrate has been found to be denoted by a slow decay of both the Ca^{2+} and PO_3^- signals and the gradual appearance of the PEEK signals, which indicated the presence of a hybrid layer where the Ca ions were embedded into the PEEK material. The literature [47] has outlined the sequence of events leading to the formation of a CaP thin film via RFMS. At the initial stages of deposition an enrichment of Ca at the substrate surface is known to take place, thought to be due to the re-sputtering of P ions, leading to a Ca/P higher than stoichiometric HA. This work has not highlighted this trend as the surface at 10 min (the HA10 sample) has been shown by XPS analysis to have already become P rich. It was thought to be the case that the investigation would have needed to have been carried out much earlier to replicate the Ca enrichment trend, as within the work completed by López et al. [48]. Evidence to conclude that Ca was indeed embedded into the PEEK material intensified as it was realised that there was less % AC of Ca-related material on the polymer surface in comparison to the thin films sputtered onto Ti, under the same conditions in previous work by the authors [3,7,20]. It is suggested that this may have been due to the heavier nature of Ca, in comparison to the P ion during bombardment, or the softer nature of the polymer substrate. This would account for the P rich surface, the lower than expected Ca/P ratio, as well as the lower than expected Ca intensity in the ToFSIMS survey scans at all time points.

Ideally, here the aim was to produce CaP coatings onto PEEK to provide a means to enhance their bioactivity. Key to this was to produce coatings that have properties commensurate with those properties outlined in the ISO 13,779 2 (2018) and ASTM F1609 standards, ideally mimicking the properties of HA, with a Ca/P ratio of around 1.67. It is clear from the results produced here that this has not been achieved as the reported Ca/P ratios are well below 1.0. The next phase of any work here would be to optimise the sputtering process to achieve enhanced coating properties to align with the ISO and ASTM standards. Aspects of substrate biasing, the process gas pressure, and the process gas used would all be important in order to achieve this. It would also be important to undertake mechanical testing of the coatings, understanding the coating thickness better, and importantly understanding the dissolution behaviour of these surfaces before progressing to more involved in vitro testing of these surfaces. If the appropriate chemical, physical, mechanical, and in vitro properties were achieved, these coatings could provide a basis for enhancing the bioactivity of PEEK materials for use in orthopaedics in order to improve bone apposition, both in terms of how quickly this can be achieved and the enhancement of the bond with bone when compared to pure PEEK devices.

5. Conclusion

From the results obtained here, it is clear that the RF magnetron sputtering has the capability to be utilised to sputter deposit, in a one-step (direct) process, homogeneous, and amorphous CaP surfaces that consist of a porous micro-/nano-structured lattice network containing appreciable levels of both Ca and P at low deposition powers (150W). It was realised that the manipulation of the deposition time had a significant influence on the chemical and physical properties of the resultant thin films, particularly in relation to the surface topography, pore area, pore diameter, and number of pores. The Ca/P of the resultant modification was low; too low in fact to coincide with HA or other bioactive

CaP phases at ~0.85. It was postulated that this was due to the bombardment of the sputtered species coming into the contact with the polymeric material, causing species to embed into the PEEK substrate, which created a matrix interlayer between the CaP material and the PEEK substrate. Further to this, the build-up of a negative charge on the insulation PEEK surface prevented re-sputtering of the P species on the surface of the surface, as would normally be observed. It was indicated that the initial 60 min of sputter deposition could be an erosion/deposition process with the capability of causing physical and chemical alterations to the surface of the underlying polymeric substrate. The next step in this work would be to investigate the mechanical properties and the dissolution behaviour of the surfaces in physiological conditions, and to determine their in vitro potential using osteoblasts.

Author Contributions: Conceptualization, A.R.B., B.J.M., and L.R.; methodology, A.R.B., J.G.A., S.H., and L.R.R.; software, L.R., S.H., J.G.A., and A.R.B.; validation, S.H., L.R., J.G.A., B.J.M., and A.R.B.; formal analysis, S.H., L.R., J.G.A., B.J.M., and A.R.B.; investigation, L.R., S.H., J.G.A., and A.R.B.; resources, A.R.B. and B.J.M.; data curation, L.R., S.H., J.G.A., and A.R.B.; writing—original draft preparation, L.R., S.H., and A.R.B.; writing—review and editing, S.H., L.R., J.G.A., B.J.M., and A.R.B.; visualization, L.R., S.H., J.G.A., and A.R.B.; supervision, A.R.B. and B.J.M.; project administration, A.R.B. and B.J.M.; funding acquisition, A.R.B. and B.J.M. All authors have read and agreed to the published version of the manuscript.

Funding: This research was funded by the Department for the Economy (Northern Ireland) and the Meehan Family Scholarship, which facilitated part of the ToF-SIMS experiments.

Acknowledgments: The authors would like to acknowledge the support of Invibio Ltd., (Thornton Cleveleys, UK), particularly Mark Brady, who kindly provided PEEK-OPTIMA™ LT1 samples for these experiments. The authors would also like to acknowledge David Scurr and Matthew Piggott at the Interface and Surface Analysis Centre, University of Nottingham for their assistance with the ToFSIMS experiments.

Conflicts of Interest: The authors declare no conflict of interest.

References

1. Rabiei, A.; Sandukas, S. Processing and evaluation of bioactive coatings on polymeric implants. *J. Biomed. Mater. Res. Part. A* **2013**, *101*, 2621–2629. [CrossRef]
2. Ha, S.W.; Kirch, M.; Birchler, F.; Eckert, K.L.; Mayer, J.; Wintermantel, E.; Sittig, C.; Pfund-Klingenfuss, I.; Textor, M.; Spencer, N.D.; et al. Surface activation of polyetheretherketone (PEEK) and formation of calcium phosphate coatings by precipitation. *J. Mater. Sci. Mater. Med.* **1997**, *8*, 683–690. [CrossRef] [PubMed]
3. Boyd, A.R.; Rutledge, L.; Randolph, L.D.; Meenan, B.J. Strontium-substituted hydroxyapatite coatings deposited via a co-deposition sputter technique. *Mater. Sci. Eng. C Mater. Biol. Appl.* **2015**, *46*, 290–300. [CrossRef] [PubMed]
4. Gloria, A.; Manto, L.; Santis, R.D.E.; Ambrosio, L. Biomechanical behavior of a novel composite intervertebral body fusion device. *J. Appl. Biomater. Biomech.* **2008**, *6*, 163–169. [PubMed]
5. Gloria, A.; Russo, T.; D'Amora, U.; Santin, M.; Santis, R.; De Ambrosio, L. Customised multiphasic nucleus/annulus scaffold for intervertebral disc repair/regeneration. *Connect. Tissue Res.* **2019**, *61*, 152–162. [CrossRef] [PubMed]
6. Duarte, R.M.; Correia-Pinto, J.L.; Reis, R.C.; Duarte, A.R.C. Advancing spinal fusion: Interbody stabilization by in situ foaming of a chemically modified polycaprolactone. *J. Tissue Eng. Regen. Med.* **2020**, *14*, 1465–1475. [CrossRef]
7. O'Kane, C.; Duffy, H.; Meenan, B.J.; Boyd, A.R. The influence of target stoichiometry on the surface properties of sputter deposited calcium phosphate thin films. *Surf. Coat. Technol.* **2008**, *203*, 121–128. [CrossRef]
8. Storrie, H.; Stupp, S.I. Cellular response to zinc-containing organoapatite: An in vitro study of proliferation, alkaline phosphatase activity and biomineralization. *Biomaterials* **2005**, *26*, 5492–5499. [CrossRef]
9. Lu, H.B.; Campbell, C.T.; Graham, D.J.; Ratner, B.D. Surface characterization of hydroxyapatite and related calcium phosphates by XPS and TOF-SIMS. *Anal. Chem.* **2000**, *72*, 2886–2894. [CrossRef]
10. Aina, V.; Bergandi, L.; Lusvardi, G.; Malavasi, G.; Imrie, F.E.; Gibson, I.R.; Cerrato, G.; Ghigo, D. Sr-containing hydroxyapatite: Morphologies of HA crystals and bioactivity on osteoblast cells. *Mater. Sci. Eng. C Mater. Biol. Appl.* **2013**, *33*, 1132–1142. [CrossRef]

11. Wolke, J.G.C.; Van Der Waerden, J.P.C.M.; De Groot, K.; Jansen, J.A. Stability of radiofrequency magnetron sputtered calcium phosphate coatings under cyclically loaded conditions. *Biomaterials* **1997**, *18*, 483–488. [CrossRef]
12. Long, J.D.; Xu, S.; Cai, J.W.; Jiang, N.; Lu, J.H.; Ostrikov, K.N.; Diong, C. Structure, bonding state and in-vitro study of Ca–P–Ti film deposited on Ti6Al4V by RF magnetron sputtering. *Mater. Sci. Eng. C* **2002**, *20*, 175–180. [CrossRef]
13. Bolbasov, E.N.; Rybachuk, M.; Golovkin, A.S.; Antonova, L.V.; Shesterikov, E.V.; Malchikhina, A.I.; Novikov, V.; Anissimov, Y.; Tverdokhlebov, S. Surface modification of poly (l-lactide) and polycaprolactone bioresorbable polymers using RF plasma discharge with sputter deposition of a hydroxyapatite target. *Mater. Lett.* **2014**, *132*, 281–284. [CrossRef]
14. Tverdokhlebov, S.I.; Bolbasov, E.N.; Shesterikov, E.V.; Antonova, L.V.; Golovkin, A.S.; Matveeva, V.G.; Petlin, D.; Anissimov, Y. Modification of polylactic acid surface using RF plasma discharge with sputter deposition of a hydroxyapatite target for increased biocompatibility. *Appl. Surf. Sci.* **2015**, *329*, 32–39. [CrossRef]
15. Tverdokhlebova, S.I.; Bolbasova, E.N.; Shesterikova, E.V.; Malchikhinaa, A.L.; Novikovb, V.A.; Anissimovc, Y.G. Research of the surface properties of the thermoplastic copolymer of vinilidene fluoride and tetrafluoroethylene modified with radio-frequency magnetron sputtering for medical application. *Appl. Surf. Sci.* **2012**, *263*, 187–194. [CrossRef]
16. Socol, G.; Macovei, A.M.; Miroiu, F.; Stefan, N.; Duta, L.; Dorcioman, G.; Mihailescu, I.; Petrescu, S.; Stan, G.; Marcov, D.; et al. Hydroxyapatite thin films synthesized by pulsed laser deposition and magnetron sputtering on PMMA substrates for medical applications. *Mater. Sci. Eng. B Solid State Mater. Adv. Technol.* **2010**, *169*, 159–168. [CrossRef]
17. Ivanova, A.A.; Surmeneva, M.A.; Tyurin, A.I.; Pirozhkova, T.S.; Shuvarin, I.A.; Prymak, O.; Epple, M.; Chaikina, M.; Surmenev, R.A. Fabrication and physico-mechanical properties of thin magnetron sputter deposited silver-containing hydroxyapatite films. *Appl. Surf. Sci.* **2016**, *360*, 929–935. [CrossRef]
18. Van Dijk, K.; Verhoeven, J.; Marée, C.H.M.; Habraken, F.H.P.M.; Jansen, J.A. Study of the influence of oxygen on the composition of thin films obtained by r.f. sputtering from a $Ca_5(PO_4)_3$ OH target. *Thin Solid Film.* **1997**, *304*, 191–195. [CrossRef]
19. Boyd, A.R.; Rutledge, L.; Randolph, L.D.; Mutreja, I.; Meenan, B.J. The deposition of strontium-substituted hydroxyapatite coatings. *J. Mater. Sci. Mater. Med.* **2015**, *26*, 65. [CrossRef]
20. Boyd, A.R.; O'Kane, C.; Meenan, B.J. Control of calcium phosphate thin film stoichiometry using multi-target sputter deposition. *Surf. Coat. Technol.* **2013**, *233*, 131–139. [CrossRef]
21. Thomas, S.; Durand, D.; Chassenieux, C.; Jyotishkumar, P. *Handbook of Biopolymer-Based Materials: From Blends and Composites to Gels and Complex Networks*; John Wiley & Sons: Hoboken, NJ, USA, 2013.
22. ASTM International. *Standard Guide to Charge Control and Charge Referencing Techniques in X-ray Photoelectron Spectroscopy*; ASTM Standard E 1523-03; ASTM International: West Conshohocken, PA, USA, 2003.
23. Dawson, P.C.; Blundell, D.J. X-ray data for poly (aryl ether ketones). *Polymer* **1980**, *21*, 577–578. [CrossRef]
24. Ha, S.W.; Hauert, R.; Ernst, K.H.; Wintermantel, E. Surface analysis of chemically-etched and plasma-treated polyetheretherketone (PEEK) for biomedical applications. *Surf. Coat. Technol.* **1997**, *96*, 293–299. [CrossRef]
25. Laurens, P.; Sadras, B.; Decobert, F.; Arefi-Khonsari, F.; Amouroux, J. Enhancement of the adhesive bonding properties of PEEK by excimer laser treatment. *Int. J. Adhes. Adhes.* **1998**, *18*, 19–27. [CrossRef]
26. Comyn, J.; Mascia, L.; Xiao, G.; Parker, B.M. Plasma-treatment of polyetheretherketone (PEEK) for adhesive bonding. Spec Issue honour Dr K.W. Allen Occas his 70th Birthd. *Int. J. Adhes. Adhes.* **1996**, *16*, 97–104. [CrossRef]
27. Henneuse-Boxus, C.; Poleunis, C.; De-Ro, A.; Adriaensen, Y.; Bertrand, P.; Marchand-Brynaert, J. Surface functionalization of PEEK films studied by time-of-flight secondary ion mass spectrometry and X-ray photoelectron spectroscopy. *Surf. Interface Anal.* **1999**, *27*, 142–152. [CrossRef]
28. Pawson, D.J.; Ameen, A.P.; Short, R.D.; Denison, P.; Jones, F.R. An investigation of the surface chemistry of poly (ether etherketone). I. The effect of oxygen plasma treatment on surface structure. *Surf. Interface Anal.* **1992**, *18*, 13–22. [CrossRef]
29. Beamson, G.; Briggs, D. *High Resolution XPS of Organic Polymers;* The scienta ESCA300 Database; John Wiley & Sons: New York, NY, USA, 1992.
30. Lub, J.; van Vroonhoven, F.C.M.; van Leyen, D.; Benninghoven, A. Static secondary ion mass spectrometry analysis of polycarbonate surfaces. Effect of structure and of surface modification on the spectra. *Polymer* **1988**, *29*, 998. [CrossRef]

31. Louette, P.; Bodino, F.; Pireaux, P.P. Poly (ether ether ketone) (PEEK) XPS reference core level and energy loss spectra. *Surf. Sci. Spectra* **2005**, *12*, 149–153. [CrossRef]
32. Capuccini, C.; Torricelli, P.; Sima, F.; Boanini, C.; Ristoscu, C.; Bracci, B.; Socol, G.; Fini, M.; Mihailescu, I.; Bigi, A. Strontium-substituted hydroxyapatite coatings synthesized by pulsed-laser deposition: In vitro osteoblast and osteoclast response. *Acta Biomater.* **2008**, *4*, 1885–1893. [CrossRef]
33. Hahn, B.D.; Park, D.S.; Choi, J.J.; Ryu, J.; Yoon, W.H.; Choi, J.H.; Kim, J.-W.; Ahn, C.-W.; Kim, H.-E.; Yoon, B.-H.; et al. Osteoconductive hydroxyapatite coated PEEK for spinal fusion surgery. *Appl. Surf. Sci.* **2013**, *283*, 6–11. [CrossRef]
34. Mutreja, I. *Controlled Dissolution of CaP Thin Films via Surface Engineering*; Ulster University: Coleraine, Ireland, 2012.
35. Almasi, D.; Izman, S.; Assadian, M.; Ghanbari, M.; Abdul Kadir, M.R. Crystalline ha coating on peek via chemical deposition. *Appl. Surf. Sci.* **2014**, *314*, 1034–1040. [CrossRef]
36. Yan, L.; Leng, Y.; Weng, L.T. Characterization of chemical inhomogeneity in plasma sprayed hydroxyapatite coatings. *Biomaterials* **2003**, *24*, 2585–2592. [CrossRef]
37. Allen, G.C.; Ciliberto, E.; Fragal'a, I.; Spoto, G. Surface and bulk study of calcium phosphate bioceramics obtained by Metal Organic Chemical Vapor Deposition. *Nucl. Instrum. Methods Phys. Res. B* **1996**, *116*, 457–460. [CrossRef]
38. Surmenev, R.A.; Surmeneva, M.A.; Grubova, I.Y.; Chernozem, R.V.; Krause, B.; Baumbach, T.; Loza, K.; Epple, M. RF magnetron sputtering of a hydroxyapatite target: A comparison study on polytetrafluorethylene and titanium substrates. *Appl. Surf. Sci.* **2017**, *414*, 335–344. [CrossRef]
39. Feddes, B.; Wolke, J.G.C.; Jansen, J.A.; Vredenberg, A.M. Radio frequency magnetron sputtering deposition of calcium phosphate coatings: Monte Carlo simulations of the deposition process and depositions through an aperture. *J. Appl. Phys.* **2003**, *93*, 662–670. [CrossRef]
40. Van Dijk, K.; Schaeken, H.; Wolke, J.; Jansen, J. Influence of annealing temperature on RF magnetron sputtered calcium phosphate coatings. *Biomaterials* **1996**, *17*, 405–410. [CrossRef]
41. Yang, Y.; Kim, K.-H.; Mauli Agrawal, C.; Ong, J.L. Effect of post-deposition heating temperature and the presence of water vapor during heat treatment on crystallinity of calcium phosphate coatings. *Biomaterials* **2003**, *24*, 5131–5137. [CrossRef]
42. Wolke, J.G.C.; De Groot, K.; Jansen, J.A. In vivo dissolution behavior of various RF magnetron sputtered Ca-P coatings. *J. Biomed. Mater. Res.* **1998**, *39*, 524–530. [CrossRef]
43. Shi, J.Z.; Chen, C.Z.; Yu, H.J.; Zhang, S.J. Application of magnetron sputtering for producing bioactive ceramic coatings on implant materials. *Bull. Mater. Sci.* **2008**, *31*, 877–884. [CrossRef]
44. Surmenev, R.A. A review of plasma-assisted methods for calcium phosphate-based coatings fabrication. *Surf. Coat. Technol.* **2012**, *206*, 2035–2056. [CrossRef]
45. Harton, S.; Stevie, F.; Ade, H. Secondary ion mass spectrometry depth profiling of amorphous polymer multilayers using O_2^+ and Cs^+ ion bombardment with a magnetic sector instrument. *J. Vac. Sci. Technol. A* **2006**, *24*, 362–368. [CrossRef]
46. Shin, M.; Brison, J.; Castner, D. ToF-SIMS depth profiling of trehalose: The Effect of analysis beam dose on the quality of depth profiles. *Surf. Interface Anal.* **2011**, *43*, 58–61.
47. Surmenev, R.A.; Surmeneva, M.A.; Evdokimov, K.E.; Pichugin, V.F.; Peitsch, T.; Epple, M. The influence of the deposition parameters on the properties of an rf-magnetron-deposited nanostructured calcium phosphate coating and a possible growth mechanism. *Surf. Coat. Technol.* **2011**, *205*, 3600–3606. [CrossRef]
48. Lopez, E.O.; Mello, A.; Farina, M.; Rossi, A.M.; Rossi, A.L. Nanoscale analysis of calcium phosphate films obtained by RF magnetron sputtering during the initial stages of deposition. *Surf. Coat. Technol.* **2015**, *279*, 16–24. [CrossRef]

Article

In Vivo Assessment of Bone Enhancement in the Case of 3D-Printed Implants Functionalized with Lithium-Doped Biological-Derived Hydroxyapatite Coatings: A Preliminary Study on Rabbits

Liviu Duta [1,*], Johny Neamtu [2], Razvan P. Melinte [2], Oana A. Zureigat [2], Gianina Popescu-Pelin [1], Diana Chioibasu [3,4], Faik N. Oktar [5,6] and Andrei C. Popescu [3,*]

1 National Institute for Lasers, Plasma and Radiation Physics, 077125 Magurele, Romania; gianina.popescu@inflpr.ro
2 Faculty of Pharmacy, University of Medicine & Pharmacy, 200349 Craiova, Romania; johny.neamtu@umfcv.ro (J.N.); razvan.melinte@gmail.com (R.P.M.); oana.toba@yahoo.com (O.A.Z.)
3 Center for Advanced Laser Technologies (CETAL), National Institute for Lasers, Plasma and Radiation Physics, 077125 Magurele, Romania; diana.chioibasu@inflpr.ro
4 Faculty of Applied Sciences, Politehnica University of Bucharest, 060042 Bucharest, Romania
5 Department of Bioengineering, Faculty of Engineering, Marmara University, Istanbul 34722, Turkey; foktar@marmara.edu.tr
6 Center of Nanotechnology & Biomaterials Application & Research, Marmara University, Istanbul 34722, Turkey
* Correspondence: liviu.duta@inflpr.ro (L.D.); andrei.popescu@inflpr.ro (A.C.P.); Tel.: +40-21-457-4550 (ext. 2023) (L.D.); +40-21-457-4550 (ext. 2423) (A.C.P.)

Received: 15 September 2020; Accepted: 15 October 2020; Published: 17 October 2020

Abstract: We report on biological-derived hydroxyapatite (HA, of animal bone origin) doped with lithium carbonate (Li-C) and phosphate (Li-P) coatings synthesized by pulsed laser deposition (PLD) onto Ti6Al4V implants, fabricated by the additive manufacturing (AM) technique. After being previously validated by in vitro cytotoxicity tests, the Li-C and Li-P coatings synthesized onto 3D Ti implants were preliminarily investigated in vivo, by insertion into rabbits' femoral condyles. The in vivo experimental model for testing the extraction force of 3D metallic implants was used for this study. After four and nine weeks of implantation, all structures were mechanically removed from bones, by tensile pull-out tests, and coatings' surfaces were investigated by scanning electron microscopy. The inferred values of the extraction force corresponding to functionalized 3D implants were compared with controls. The obtained results demonstrated significant and highly significant improvement of functionalized implants' attachment to bone (p-values $\leq$0.05 and $\leq$0.00001), with respect to controls. The correct placement and a good integration of all 3D-printed Ti implants into the surrounding bone was demonstrated by performing computed tomography scans. This is the first report in the dedicated literature on the in vivo assessment of Li-C and Li-P coatings synthesized by PLD onto Ti implants fabricated by the AM technique. Their improved mechanical characteristics, along with a low fabrication cost from natural, sustainable resources, should recommend lithium-doped biological-derived materials as viable substitutes of synthetic HA for the fabrication of a new generation of metallic implant coatings.

Keywords: biological-derived hydroxyapatite coatings; lithium doping; food industrial by-products; in vivo extraction force; pulsed laser deposition; 3D printing

---

1. Introduction

The increase of life expectancy and the enhanced frequency of injuries and diseases are considered the most important causes for the escalating demand for dental and orthopedic devices. In this respect, the surface functionalization of implants with highly performant bioactive materials is currently of interest and necessary both for the prevention of failure and the prolongation of the bone implants' life. Thus, good examples are calcium phosphates (CaPs) and bioglasses [1–4], which are the most used bioceramic materials in medicine, in particular in orthopedics and dentistry [5,6], as coatings for various metallic implants [7]. From the first class, a special focus was put on hydroxyapatite (HA), $Ca_{10}(PO_4)_6(OH)_2$ [3,4,7]. Due to their role as scaffolds for osteogenic differentiation [8], the ability to form strong bonds with the host bone tissues [9], and their excellent capacity to stimulate and accelerate the formation of new bone tissue around implants [10–12], HA ceramics are frequently utilized in bone grafting and dental devices as bone substitutes, either in their simple form, or in conjunction with other different biomaterials [13,14]. One should note that, if for the healthy patients, the osseointegration rate takes place at a reasonable speed, in the case of older patients, or those presenting critical medical conditions from the systemic point of view (in particular HIV positive patients that reported an increase in early failures of dental implants [15,16]), this process could be improved by applying HA-based coatings on to the implants' surfaces.

Despite its excellent bone-regeneration properties, HA is brittle in bulk [17] and characterized by poor mechanical properties. Usually, an implant can be manufactured from Ti or its medical-grade alloys. In this respect, additive manufacturing (AM) is a technology that allows for cost-effective and rapid production of complex three dimensional (3D) metallic parts and is gaining nowadays increased attention in the field of personalized medicine [18,19]. In contrast to their excellent mechanical properties, Ti implants elicit low osseointegration rates. To overpass this shortcoming, HA can be applied as a coating onto the surface of Ti implants, to significantly improve the overall performances of the structures, by successfully combining the excellent bioactivity of the ceramic with the mechanical advantages of the metallic substrates [17,20,21]. To enhance these properties even more, HA doping with various concentrations of therapeutic ions is envisaged [4,7,22,23].

Currently, huge interest is focused on finding appropriate deposition techniques to modify the surface of metallic implants, by functionalization with coatings, proteins and/or drugs that could favor the enhancement of cellular adhesion, thus leading to the acceleration of the osseointegration time [24–27]. In the field of thin-film growth, pulsed laser deposition (PLD) has been established in the past 15–20 years as one of the most popular and efficient methods for the deposition of a wide spectrum of materials, especially onto substrates with complex geometrical shapes, including implants [28]. Moreover, when CaPs of animal origin are involved, the PLD technique could be one of the most suitable choices to synthesize this type of coating. If the synthetic HA has a complex stoichiometry, with a large number of atoms and functional groups difficult to be transferred in the form of thin layers by physical vapor deposition techniques, biological apatites become even more difficult to transfer due to the supplemental presence of functional groups and substitutional ions, which further complicate their stoichiometry and structure [29]. Nevertheless, the PLD method is well-known for its ability to stoichiometrically transfer very complex molecules (due to the high ablation rate which causes all elements to evaporate at the same time [30]) and this advantage should place it among the main candidates for the successful transfer of such complex materials. Moreover, with the major drawback of this deposition technique (i.e., being capable of coating only small-area substrates, as the plasma plume diameter is restricted by the low spot size of the laser beam) being overcome by the implementation of laser scanning units in commercially, widely available PLD equipment, the path towards large-scale application is thus opened [31,32].

It is a documented fact that in vitro tests can play an important role in the evaluation of a material by providing useful information about its potential behavior in a biological environment. To better understand the complex processes that may occur in a living system and to provide the most accurate data for the complete confirmation of the performance of a biomaterial designed for clinical trials, in vivo tests in animal models are extremely important and should, therefore, follow thorough in vitro confirmation. Accordingly, after previous validation of both in vitro cytocompatibility [33,34] and antimicrobial activity [35] of biological HA (BHA) doped with lithium carbonate (further denoted as Li-C) and phosphate (further denoted as Li-P) coatings, a leap forward in our research was performed by investigating their preliminary bone bonding and bone bonding strength characteristics in vivo in an animal model. Thus, the aim of this preliminary study was both to (i) demonstrate superior bone–implant interactions in the case of PLD functionalized implants as compared to simple, uncoated ones (Ti controls), and (ii) verify the hypothesis that higher values of the extraction force are obtained in the case of implants extracted after longer periods of implantation time. It should be stressed that there are reports in the literature on the in vivo testing of CaPs (especially HA), but, to the best of our knowledge, the novelty of the current work resides both in the additive manufacturing (AM) technique used to fabricate the implantable medical devices, and the fact that this is the first preliminary study to address the in vivo mechanical behavior of these 3D printed implants functionalized by PLD with biological HA, doped with Li-C and Li-P coatings.

2. Materials and Methods

2.1. Printing of Metallic Implants

The laser melting deposition (LMD) technique was used for the manufacturing of three-dimensional (3D) metallic implants. A Ti6Al4V powder (with the particle size of <90 µm), further denoted as control, was used as precursor material. Implants were in "T"-shape (Figure 1a,b) and their dimensions are displayed in Figure 1a. It should be mentioned that, by using the LMD technique, the fabrication costs and manufacturing time of 3D metallic implants were significantly reduced.

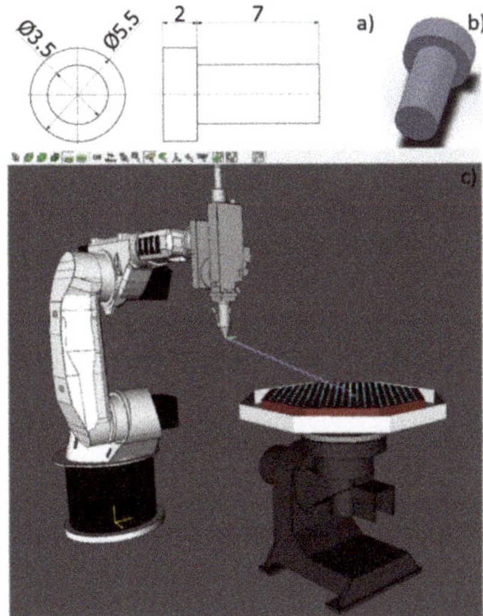

Figure 1. (**a**) The technical drawing of the 3D metallic implant; (**b**) the 3D drawing of the implant; (**c**) the TruTops Cell® graphical software which generates the movement codes for the robotic arm.

First, incipient bulk prisms of $(10 \times 5 \times 1)$ cm^3 were 3D printed using several experimental conditions, to find the optimal parameters in view of obtaining defect-free samples. Thus, by optimization of the laser and scanning parameters, compact structures without cracks or pores and with excellent compositional uniformity were obtained. The laser beam (supplied through an optical fiber) used for melting the powder was generated by an Yb:YAG disk source (λ = 1030 nm, continuous wave), model TruDisk 3001 (Trumpf, Ditzingen, Germany). The surface focused spot size had a diameter of ~0.6 mm.

An automated powder feeder (Trumpf, Ditzingen, Germany), connected through hoses (having 6 mm diameters) to a nozzle with three flow channels, that was mounted on a robotic system TruLaser Robot 5020 (Trumpf, Ditzingen, Germany), made of a Kr30HA robot (Kuka, Augsburg, Germany), with six movement axes, was used for the powder delivery. The powder was sent to the robot through a delivery system provided with turntables. A three-beam particle flow nozzle was used and aligned so that the beams could be coherent in a single point, congruent with the laser beam spot. The Ti implants were fabricated under Ar ambient.

The 2D technical drawing of the 3D metallic implants (Figure 1a) was generated in the graphical engineering software Solid Works® (Dassault Systems, Vélizy-Villacoublay, France), then translated to the 3D space (Figure 1b), and subsequently imported in TruTops Cell® (Trumpf, Ditzingen, Germany) (Figure 1c), a movement code generator for the robotic arm.

The design of the trajectory followed by the laser beam for structure growth proved to be essential to obtain samples that respect both the appearance and dimensions of the technical drawings. The separation distance of contours on the *z*-axis, established by tracing a calibration curve, was of 0.5 mm. All trajectories executed by the robotic arm were extremely important to obtain a homogeneous structure, free of discontinuities.

Three-dimensional metallic implants were fabricated onto a Ti substrate with a thickness of 10 mm, after which they were removed from it by a disk cutting machine model Brillant 200 (ATM, Mammelzen, Germany).

2.2. *Pulsed Laser Deposition (PLD) Experiments*

2.2.1. Powders

Biological-derived hydroxyapatite (BHA) powders were obtained from the cortical part of bovine femoral bones, according to the protocol described in [29]. Concisely, heads of femoral bones were cut-off and shafts were further processed. The bone marrows were then extracted and the unwanted soft tissue residues or macroscopic adhering impurities and substances were gently removed from shafts, which were sliced, cleaned and washed with distilled water and deproteinized for 14 days in an alkali media of 1% sodium hypochlorite. The elimination of any potential hazardous contaminants was achieved by a calcination process (at 850 °C, for 4 h, in air) of the resulting dry bone fragments [36,37]. Prior to undergoing ball-milling to fine powders, the calcined bone specimens were crushed with a mortar and pestle. It should be emphasized that, for the fabrication of BHA powders, European Union (EU) regulations [38] and ISO 22442-1 [39] were followed. Batches of BHA fine powders were admixed with 1 wt.% of Li-C and Li-P (Sigma-Aldrich GmbH, St. Louis, MO, USA).

2.2.2. Target Preparation

The as-obtained Li-C and Li-P mixed powders were pressed at ~6 MPa in a 20 mm diameter mold. The resulting pellets were thermally treated in air, using an oven, for 4 h, at 700 °C. A heating rate of 20 °C/min and a cooling ramp of 5 °C/min were applied. Following this protocol, the fabrication of hard and compact targets was carried out.

2.2.3. Coating Fabrication

PLD experiments were conducted inside a stainless-steel deposition chamber, in an ambient water vapor pressure of 50 Pa. The target-to-substrate separation distance was of 5 cm. Coatings were synthesized using a KrF* excimer laser source (COMPexPro 205, Coherent, Santa Clara, CA, USA, λ = 248 nm, $\tau_{FWHM} \leq$ 25 ns). The incident laser fluence was set at 3.5 J/cm^2 (with a corresponding pulse energy of 360 mJ). The laser beam was incident at 45° on the target surface. For the growth of one film, 15000 consecutive laser pulses were applied. During the multi-pulse laser irradiation, the target was continuously rotated with 0.3 Hz and translated along two orthogonal axes, to avoid piercing and to obtain unidirectional plasma.

Prior to introduction into the deposition chamber, all substrates were successively cleaned following a three-step protocol [34]. In addition, to avoid any possible micro-contamination, targets were submitted to a "cleaning" process with 1000 laser pulses. To collect the flux of expulsed micro-impurities, a shutter was interposed between the target and the 3D metallic substrates.

To ensure the complete deposition of the 3D Ti implants, a rotating stainless-steel support flange and four clamping pins [40] were used. During depositions, the substrates' temperature was kept constant at (500 ± 5%) °C, using a heater connected to a PID-EXCEL temperature controller (EXCEL Instruments, Mumbai, India). A heating rate of 25 °C/min and a cooling ramp of 10 °C/min were applied.

2.2.4. Thermal Treatments

Both the temperature used for heating the substrates during experiments and post-deposition thermal treatments, were demonstrated to have an important influence on the coatings' characteristics and, in particular, on their composition and stoichiometry. Therefore, all synthesized structures were submitted to post-deposition thermal-treatments in water-vapors enriched atmosphere. A removable flange, positioned on the heater so as to ensure an optimal contact with the substrates, was used.

2.3. Animals and Surgical Experimental Protocol

The animal spectrum generally used to test CaP coatings synthesized by the PLD technique is limited to rats, mini-pigs, dogs, goats, sheep and rabbits [17]. Both the advantages and disadvantages of using different animal models have been widely discussed in the study reported by Lu et al. [41]. It should be emphasized that, the rabbit represents one of the most used animals for medical studies [42], due to its size and ease of handling. Moreover, its skeleton reaches maturity in a rather short period of time (approximatively six months) [43].

For the in vivo experiments reported in this study, that took place in the biobase of the University of Medicine and Pharmacy (UMF), Craiova, Romania, a total of 26 skeletally-matured New Zealand White rabbits, aged six months and weighing between 3 and 3.5 kg were used. One should note that, animals were weighed at the beginning of the study and before sacrifice, and no significant weight changes were observed. All animals were free of disease. They were housed in individual plastic cages, in a climate-controlled environment at 22 °C, 45% humidity, and 12-h alternating light–dark cycles. During the experimental period, free access to a standard laboratory diet and tap water were provided. All animals were acclimatized for 10 days before use in this study.

Before surgery, the control and functionalized 3D Ti implants were sterilized by autoclaving (at 120 °C, for 1 h).

The 26 rabbits were randomly assigned to two groups (n = 13 each). The 3D metallic implants were introduced into the femoral condyles [44] (two implants in each rabbit), using the following sequence: at the level of the right femur, the 3D Ti implants functionalized with Li-C and/or Li-P coatings, and at the level of the left femur, the simple Ti implants (controls).

Antibiotic prophylaxis was performed pre-operatively, by intramuscular administration of a cephalosporin (sulfate diluted in physiological serum, at a dose of 40 mg/kg body). The antibiotic treatment continued for up to seven days (two doses/day).

All surgeries were performed under general anesthesia, following a three-step procedure: (i) sedation by subcutaneous administration of fentanyl (0.1 mL/Kg) and midazolam (2 mg/Kg), (ii) maintenance (during the surgical act) of the anesthesia by administration of diluted fentanyl with physiological serum (1 mL fentanyl to 9 mL physiological serum), and (iii) injection at the incision site (under sterile conditions), of 1% xylin (5 mL), as an adjuvant.

At the beginning of the surgical procedure, the incision site was shaved and washed well with water and soap, and disinfected with betadine solution, followed by covering the animal with sterile overlays. An incision of ~3 cm was performed on the lateral face of the distal femoral epiphysis. This interested the epidermis, the dermis, and the facial layers. The muscle present at this level was dissociated and the periosteum was incised longitudinally, exposing the bony lateral face of the distal femoral epiphysis.

Using a dental burr (Stryker Core Reamer), with adjustable rotational drill speed, under continuous saline irrigation, to which burr drills (Osstem surgical kit for dental implants, model TS III SA) with progressive diameter were attached, an opening hole was drilled through the cortical bone. Later, this hole was enlarged progressively (to avoid a possible cortical fracture), to a diameter slightly inferior to the implant's dimensions. Into the bone socket thus created, the 3D metallic implant was inserted in a press-fit manner, ensuring both complete coverage of its surface and stability. The implant was then carefully covered and protected by soft tissues and the muscle fascia and epidermis were closed with sutures (Figure 2).

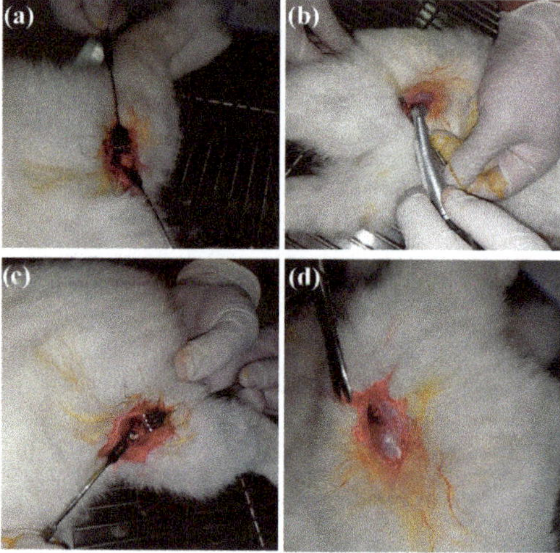

Figure 2. Surgical steps for the correct positioning of the 3D Ti implant: (**a**) exposure of lateral side of the distal femur, with visualization of the cortical bone; (**b**) opening hole drilled through the cortical bone using a dental burr; (**c**) insertion of the implant into the bone socket in a press-fit manner; (**d**) cover and protection of the implant by soft tissues and closing of the wound.

Immediate postoperative radiological examinations by computed tomography (CT) were performed both to confirm the correct position of the implants and to rule out any bone fractures or other possible surgical complications.

After the surgery, all animals were allowed to move freely in their cages, being monitored for general health conditions, on a daily basis. The operative wound was carefully checked and bandaged until complete healing. At the end of the implantation procedures, all rabbits were permitted free access to normal pellet food and water. No post-operative complications were reported.

The surgical protocol used for introducing the 3D metallic implants into the rabbits' femoral condyles was approved by the "Committee of Ethics and Academic and Scientific Deontology" at the UMF in Craiova, Romania (document no. 135/20.12.2019), and the entire experimental process and the surgical technique complied with regulations and precautions of the EU Council Directive of 22 September 2010, regarding the care and use of laboratory animals for scientific purposes (2010/63/EU).

2.4. Characterization of Control and Functionalized Three-Dimensional (3D) Ti Implants

2.4.1. Computed Tomography

Investigations were performed using a Siemens CT scanner, operated at 130 kV, 90 mA, 0.5 mm section thickness and 0.3 mm section increment. The integration and total scan times were of 0.5 s and 2 min, respectively. For the evaluation of bone density, the Onis 2.3.5 software was used, and the inferred values were expressed on the Hounsfield tissue density scale (HU units). The area on the CT sections in which the tissue density was evaluated (also known as the region of interest), was always selected from the same region of the implant, for each performed measurement.

2.4.2. Mechanical Testing

Four weeks after the insertion of the 3D metallic implants, the rabbits were anesthetized (following the same protocol described above), and euthanized by intracardiac injection, using an overdose of sodium pentobarbital (100 mg/kg).

The anterior incision was resumed and the distal femoral epiphysis, containing the implant after disarticulation and proximal sectioning of the femoral diaphysis (Figure 3), was harvested for evaluation. After careful dissection and cleaning of any adherent soft tissues, the bone tissue–implant block specimens were obtained and immediately fixed in 5% buffered formaldehyde solution, prior to being subjected to mechanical tests (within 1 h of euthanasia).

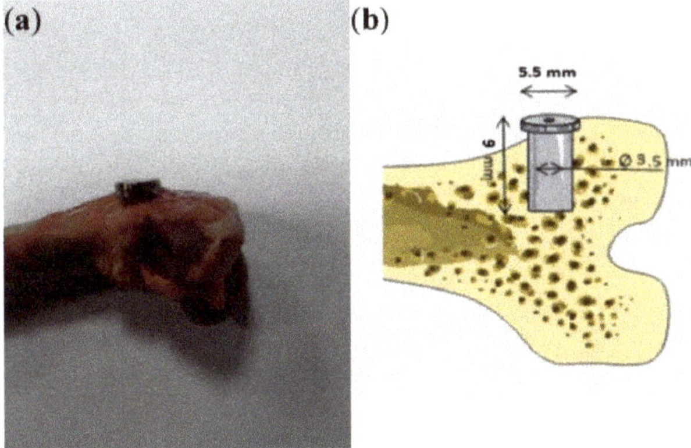

Figure 3. (**a**) Photograph of the extracted bone; (**b**) schematic representation of the implant site.

The quantification of the in vivo interfacial extraction force of bone implants represents a biological challenge. One should note that, the analysis of the phenomena that occur at the bone tissue–implant interface is a complex procedure, which requires the use of experimental cutting and grinding techniques, which are quite complicated and expensive. To avoid these shortcomings, the experimental model of in vivo testing of the extraction force of 3D metallic implants was, therefore, applied in this study.

It is important to mention that, the retention between the implant and the bone was evaluated before euthanasia could influence the results (within the first 4 h after euthanasia).

The measurement of the implants extraction force was performed by a tensile traction machine (model WDW, Time Group), which measures both the force and the elongation, and can operate with a maximum force of 5000 N. All tests were performed with a traction speed of 1 mm/min. The implant was positioned into the traction machine support by means of a hexagonal adapter attached with an adhesive on the exterior side of the 3D Ti implant (Figure 4). One should take into account that, when fixing the femur and the implant, the position of the 3D metallic implant should be aligned with the traction direction. Typical load-displacement curves were recorded (data not shown here), and the failure load was defined as the peak load value of the load-displacement curve.

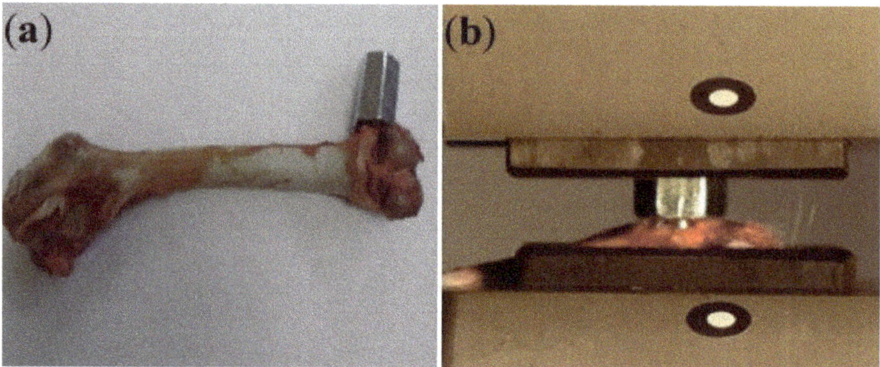

Figure 4. Tensile pull-out measurement: photographs of (**a**) the hexagonal adaptor attached to the 3D Ti implant, and (**b**) both the femur and the adaptor attached to the 3D metallic implant, in the traction machine support.

2.4.3. Scanning Electron Microscopy (SEM)

The morphological analysis of the 3D metallic implants' surfaces, following their extraction from the rabbits' femoral condyles, was performed. Therefore, to identify the adherent bone components, surfaces of both control and functionalized 3D Ti implants were examined by scanning electron microscopy (SEM). Using the analysis of backscattering electrons, along with the ImageJ free software (version 1.46r), one could estimate the adherence ratio of the bone tissue remaining on the implant's surface after its extraction from the rabbits' femoral condyle. This parameter is defined as the ratio between the estimated area of the remaining bone tissue and the area of the 3D implant (the area of a circle with a diameter of 3.5 mm).

A SEM SU5000 (Hitachi, Tokyo, Japan), with a resolution of 1.2 nm and an acceleration voltage of 25 kV, was used for these investigations. For comparison reasons only, SEM micrographs of control Ti and Li-C and Li-P structures before surgery are also presented.

2.5. Statistical Analysis

Experiments were carried out both in decuplicate ($n = 10$, for extractions performed at four weeks), and triplicate ($n = 3$, for extractions performed at nine weeks), to achieve statistical significance. The unpaired Student's t-test was used to determine the level of significance, and p-values ≤ 0.05, ≤ 0.0001 and ≤ 0.00001 were considered to be significant and highly significant, respectively.

3. Results

3.1. Clinical Observations Following Implantation

Four weeks after the implantation procedure, the performed CT scans demonstrated the correct placement and a good integration of all implants into the surrounding bone (Figure 5).

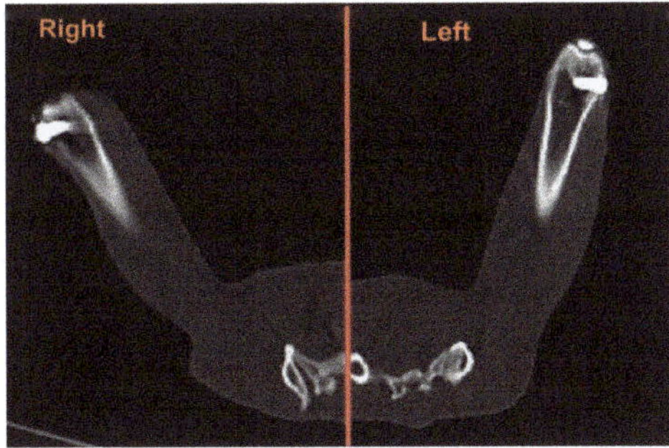

Figure 5. Lateral computed tomography images showing the positioning of the 3D metallic implants (right foot—implant functionalized with Li-C coating, left foot—control Ti implant), at 4 weeks after surgery.

The results of the osseous density measurements at the implant–bone interface, for control and functionalized (with Li-C and Li-P coatings) 3D Ti implants, are summarized in Tables 1 and 2.

Table 1. Osseous density (mean ± standard deviation, SD) inferred at the implant–bone interface for control and functionalized (with Li-C coatings) 3D Ti implants.

Implant Type	Osseous Density (Mean ± SD) [HU]	
	4 Weeks	9 Weeks
Control (Ti)	811 ± 21	850 ± 57
Li-C	1068 ± 70	1156 ± 40

Table 2. Osseous density (mean ± SD) inferred at the implant–bone interface for control and functionalized (with Li-P coatings) 3D Ti implants.

Implant Type	Osseous Density (Mean ± SD) [HU]	
	4 Weeks	9 Weeks
Control (Ti)	818 ± 36	855 ± 61
Li-P	1053 ± 40	1172 ± 28

For all investigated 3D Ti implants, CT scans revealed at the implant–bone interface the presence of the peripheral osteosclerosis, and no inflammatory process of the soft tissues.

The results presented in Tables 1 and 2 point out to an increase of the osseous density, for both investigated time periods. Therefore, the bone density values corresponding to the 3D Ti implants functionalized with both Li-C and Li-P coatings, measured at 9 weeks, were ~1.2 times higher than those inferred at 4 weeks after surgery, respectively. At 4 weeks after surgery, both functionalized 3D Ti implants (with Li-C and Li-P coatings) showed bone density values ~1.3 times higher than those obtained in the case of control 3D Ti implants. Moreover, at 9 weeks, the density values inferred in the case of functionalized 3D Ti implants were ~1.4 times higher as compared to control ones.

One should also stress upon that, all rabbits were able to walk normally within 6 h after the implantation surgery. Moreover, no macroscopic signs of infection or adverse reactions were observed, and none of the rabbits died or suffered a bone fracture during this study.

3.2. Mechanical Testing

Mechanical tests were used to evaluate the quality of the implants' osseointegration. It is important to note that no implant showed surface alteration or disruption.

In Figures 6 and 7 are represented the characteristic values of the load to failure (detachment force, F_{max}) of implants under tensile pull-out testing, inferred for control and functionalized 3D metallic implants, at 4 and 9 weeks after surgery, respectively.

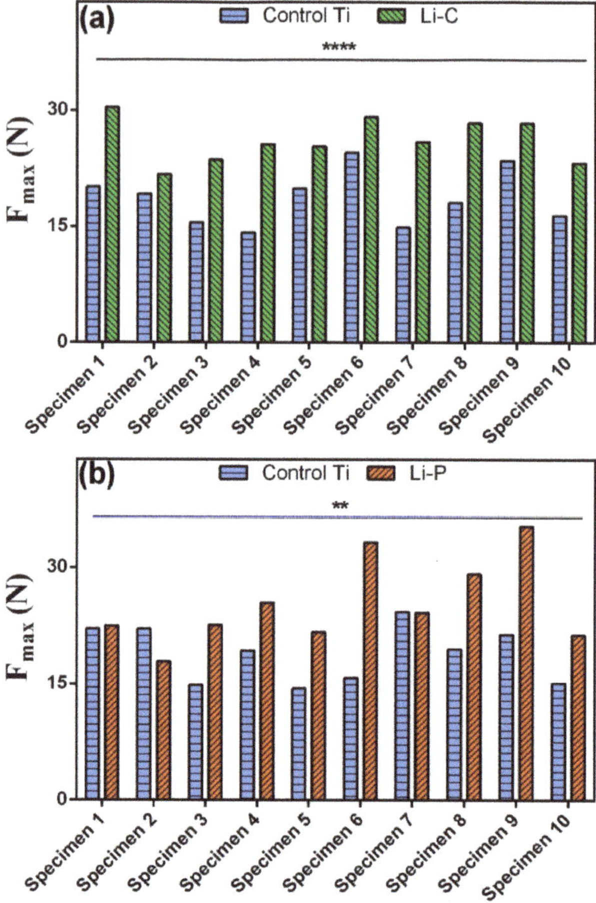

Figure 6. Detachment force, F_{max}, of implants (n = 10) under tensile pull-out testing, inferred in the case of control 3D Ti implants (marked in blue color) and of those functionalized with (**a**) Li-C (marked in green color) and (**b**) Li-P (marked in orange color) coatings, at 4 weeks after surgery. **** Represents highly significant differences ($p \leq 0.0001$). ** Represents significant differences ($p \leq 0.01$).

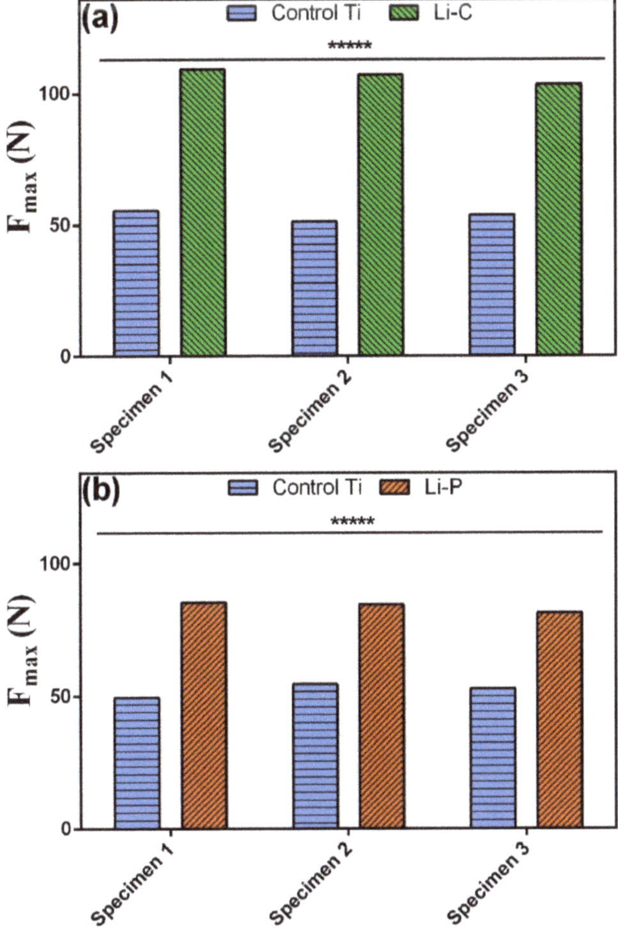

Figure 7. Detachment force, F_{max}, of implants ($n = 3$) under tensile pull-out testing, inferred in the case of control 3D Ti implants (marked in blue color) and of those functionalized with (**a**) Li-C (marked in green color) and (**b**) Li-P (marked in orange color) coatings, at 9 weeks after surgery. ***** Represents highly significant differences ($p \leq 0.00001$).

In the case of the extraction tests performed at 4 weeks (Figure 6a,b), the obtained mean detachment force values demonstrated (i) a highly significant difference between 18.6 N (±3.5), for the first control Ti group ($n = 10$), and 26.2 N (±2.9), for the Li-C test group ($n = 10$) ($p = 0.00006$) (Figure 6a), and (ii) a significant difference between 18.8 N (±3.5), for the second control Ti group ($n = 10$), and 25.4 N (±5.6), for the Li-P test group ($n = 10$) ($p = 0.007$) (Figure 6b).

When referring to the extractions performed at longer time periods, i.e., 9 weeks from surgery (Figure 7a,b), the inferred mean detachment force values indicated highly significant differences for both investigated cases: between 53.5 N (±2.2), for the first control Ti group ($n = 3$), and 106.6 N (±2.9), for the Li-C test group ($n = 3$) ($p = 0.00003$) (Figure 7a), and between 52.1 N (±2.5), for the second control Ti group ($n = 3$), and 83.7 N (±2.1), for the Li-P test group ($n = 3$) ($p = 0.00009$) (Figure 7b).

One can observe that, the failure loads of 3D Ti implants functionalized with both Li-C and Li-P coatings measured at 9 weeks were (3.4–5) and (2.3–4.8) times higher than those inferred at 4 weeks after surgery, respectively (Figures 6 and 7). It should be stressed here that, a similar trend was also observed by Yan et al., in the case of strontium-containing HA coatings [45]. Moreover, at 4 weeks after surgery, the 3D Ti implants functionalized with Li-C and Li-P coatings showed a bone attachment strength of about (1.1–1.8) and (1.1–2.1) times stronger than that corresponding to the control 3D Ti implants, respectively. After 9 weeks of implantation, the inferred values of the attachment force were about (1.9–2.1) and (1.5–1.7) times higher than controls, in the case of Li-C and Li-P samples, respectively. Therefore, one could indicate that both the PLD surface functionalization of 3D Ti implants and a longer implantation time period could positively influence the overall bone bonding strength characteristics of the investigated medical devices. With this result, the aim of this study was attained.

3.3. SEM

Immediately after deposition, the surfaces of control and functionalized 3D Ti implants were examined by SEM, under two different magnifications (1000× and 10,000×, respectively). Typical SEM micrographs of control 3D Ti implants (Figure 8) show irregular morphologies which are due to the micro-machining preparation process only. The deposition of Li-C and Li-P coatings produced surfaces with rough morphologies, made of spheroidal formations (Figure 8), known in the literature as particulates. It should be stressed here that, their origin and dependence on the target composition represent common features of the PLD process [46].

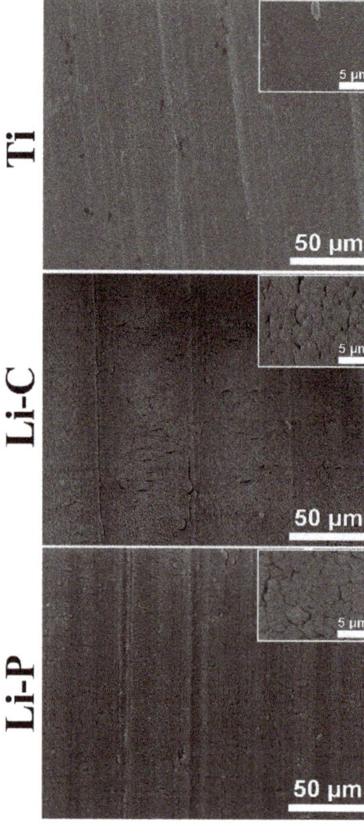

Figure 8. Scanning electron microscopy (SEM) micrographs of control and functionalized (with Li-C and Li-P coatings) 3D Ti implants before implantation.

After the extraction procedure, SEM examinations of control and functionalized 3D Ti implants were carried out under two magnifications, i.e., 300× and 2000×, respectively (Figure 9).

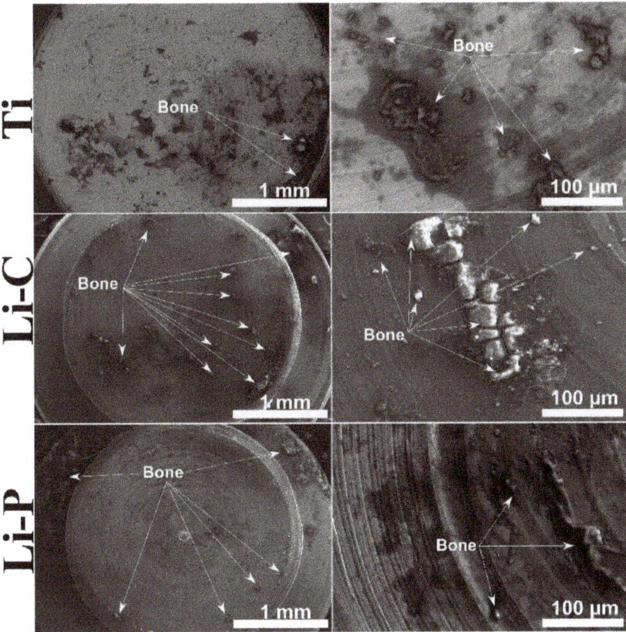

Figure 9. SEM micrographs indicating bone detachment on the surface of a control and functionalized (with Li-C and Li-P coatings) Ti implant, at 4 weeks after surgery.

The inferred adherence ratio of the remaining bone fragments onto the surface of the extracted implants revealed values of 2.69 ± 0.03%, in the case of control 3D Ti implants, and of 3.54 ± 0.03% and 3.66 ± 0.02%, in the case of the implants functionalized with Li-P and Li-C coatings, respectively. This corresponds to adherence ratios up to ~38% higher in the case of functionalized 3D Ti implants as compared to control ones (Figure 10). This could be indicative of an enhanced osseointegration process. Moreover, the higher values of the detachment force inferred in the case of functionalized 3D Ti implants, in comparison to control ones (presented in Figures 6 and 7), should support this observation. It is important to stress that, the presence of such osseous structures onto the surface of 3D Ti implants suggests, besides the beginning of the implant integration process into the bone, the absence of any adverse reactions at the implantation site.

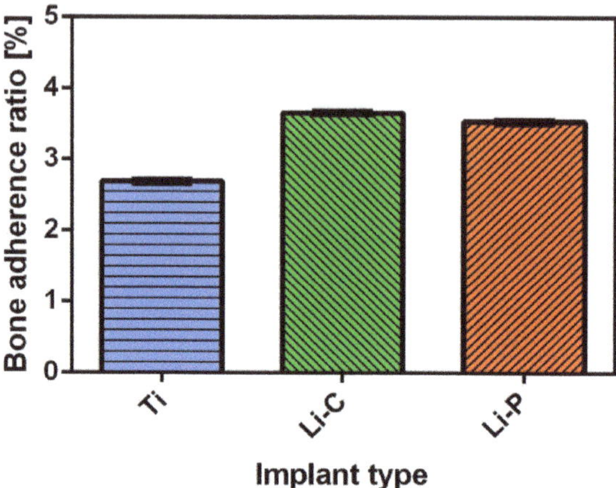

Figure 10. Inferred bone adherence ratio in the case of control and functionalized (with Li-C and Li-P coatings) 3D Ti implants (n = 3), at 4 weeks after surgery.

4. Discussion

The rabbit's bones manifest faster changes and turnover in comparison to larger animal models [47]. Taking into consideration the difficulty to extrapolate the results obtained on rabbit bones to the human ones, the current study represents a screening for implant design and a validation of the used biomaterials, before their testing on larger animal models.

Due to its attractive biological properties and resemblance to the mineral part of the bone, HA currently represents one of the most used CaPs. To overcome the drawback that synthetic HA does not completely match the chemical composition of bone mineral component [48], continuous efforts have been dedicated to find alternative, low-cost methods to produce HA, such as obtaining it from sustainable, biological (BioHA) resources (biogenic, mammalian, and/or natural fish bones). One should note that BioHA is a carbonated, non-stoichiometric Ca-deficient material, which is different from synthetic HA in terms of composition, stoichiometry, crystal size/morphology, crystallinity degree, degradation rate, and overall biological performance [29]. Another important aspect of the growing interest of using BioHA materials is related to the presence of trace elements and functional groups, which modify the chemical formula of the natural HA in bone. While the chemically-synthesized HA lacks these elements, mammalian bones contain a higher source of ions and trace elements [49–51], with Na^+ and Mg^{2+} as the most frequently found ones [29]. One notes that, the presence of these ions alongside HA plays an important role in the development of teeth and bone, whereas their absence could cause fragility or bone loss [6].

The three main steps for bone formation are (i) cellular osteogenic differentiation, (ii) matrix maturation and (iii) matrix mineralization. The initial adhesion of cells to an implant represents a key parameter for their proliferation and differentiation onto the surface of a biomaterial. The surface functionalization of implants by CaP coatings can ensure new bone formation, which might occur rapidly along the entire surface of the coating, as compared to the simple Ti implant (control) [52]. In this respect, in our previous studies [33,34], the effect of lithium addition (Li_2CO_3 and/or Li_3PO_4) in BHA coatings synthesized by PLD was thoroughly examined from the physical-chemical, mechanical and cytocompatibility points of view. In comparison to bare Ti implants and simple (undoped) BHA coatings, Li-C and Li-P structures elicited improved wettability properties, which could further induce improved bone regeneration [33]. These characteristics seemed to have a beneficial influence on the overall cytotoxicity of the materials, the proliferation of human mesenchymal stem cells (hMSC) reaching its highest values in the case of Li-C coatings, followed by Li-P and simple (undoped) BHA ones. After 72 h, the cellular viability of hMSC was superior for BHA-doped structures (Li-P and Li-C) as compared to simple (undoped) BHA ones. One should note that, the bonding strength values were two times (in the case of Li-P coatings) and up to more than three times (in the case of Li-C coatings) higher than the threshold (>15 MPa) imposed by the ISO standard regulating the load-bearing implant coatings [33]. The evaluation of simple (undoped) and doped (Li-P) BHA coatings to induce osteoblast (OB) cells differentiation was also performed [34]. The level of mineralization in the case of Li-P coatings was found to be higher and significantly higher as compared to simple (undoped) BHA ones and bare Ti, respectively. In addition, when referring to the in situ level of alkaline phosphatase (which is an early marker of osteogenic differentiation), in hMSC and differentiated OB cells, the slightly improved results in the case of doped BHA (Li-P) structures in comparison to simple (undoped) BHA ones, could be explained by the characteristics of the coatings induced by the lithium doping agent (low roughness values and a hydrophilic behavior) [34]. Moreover, the comparative morphological investigations evidenced the presence of numerous nodules of mineralization originating from OB cells grown on the surface of Li-P coatings, as well as a network-like organization of OBs interconnected into the mineralized extracellular matrix. Therefore, it was suggested that this network of OBs could constitute the premise for in vitro early differentiation of cells and, consequently, for an improved osseointegration in vivo [34]. Analyzing the overall demonstrated improved characteristics of lithium-doped BHA structures in comparison to bare Ti implants and simple (undoped) BHA coatings, it was concluded that the incorporation of lithium might prove to be a delivery vehicle for bioactive agents to promote and even accelerate osseointegration in close relation with an improved anchorage of bone metallic implants. It should be emphasized here that, the current EU strategies applied for the optimization of in vivo experiments indicate that the maximum amount of relevant, high-quality data should be generated strictly in accordance with the 3*R*s philosophy, i.e., reduction, refinement and replacement of animal use, as defined by Russell and Burch in 1959. Taking into consideration all these aspects, along with the fact that the aim of this pilot study was not to investigate the lithium effectiveness in in vivo experiments, but to demonstrate the superiority of lithium-doped BHA coatings over commercial Ti implants (which represent the "gold standard" for implantological applications), the use of simple (undoped) BHA coatings as a second control (next to bare Ti implants) was not considered necessary at this stage of research.

We have shown in the current study that, after implantation of the 3D printed coated and uncoated Ti implants, the bone in the vicinity of the devices started to react to the foreign body. In all cases, between 4 and 9 weeks the bone density increased. There was an obvious difference between the uncoated (control) 3D Ti implants and those functionalized with Li-C and Li-P coatings, in terms of bone density. The coated implants were always surrounded by a denser bone as compared to the uncoated ones, after 4 and 9 weeks, respectively. The denser bone around Li-C and Li-P coatings was expected, as the synthesized structures are based on natural HA, which is very similar to the mineral part of the bone. The material is biomimetic and bioactive, shielding the metallic implant and further stimulating the bone growth.

Apparently, between the Li-C and Li-P coated implants there were no obvious differences in terms of bone density. Therefore, to elucidate which of the two type of coatings could be more beneficial for the bone growth, in vivo mechanical tests were performed. It should be noted that, the optimal functioning of an endosseous implant depends on the mechanical stability of the biomaterial, which can be evaluated by extraction tests. To obtain information regarding the force that occurs between the bone tissue and implanted materials, various experimental study models have been developed, each of these approaches having its own particularities [44,47,53]. Most animal models currently used to study the retention of implants in bone are influenced by shear forces introduced during the retention tests. Push-out and/or torsion tests are generally used to analyze these forces. Push-out tests have some limitations regarding the evaluation of the bone–implant force. Therefore, the implant's shape, the degree of surface roughness and the good alignment between the implant clamping system and the pushing device can significantly influence test results. The use of the torsion test has the theoretical advantage of achieving an equal distribution of the force field at the bone–implant interface.

On the other hand, these experiments are far more complicated and time-consuming as compared to the push-out ones. At the same time, the torsion test does not provide a clear distinction between the chemical bone–implant interaction and the mechanical retention because of the implant's surface roughness. Therefore, some studies, including the current one, have been concentrated on tensile strength measurements which, very important, can be influenced only by the chemical links between the implant and the bone [54–56]. Therefore, it is considered that by tensile tests the strength of the chemical bond can be measured directly. Thus, after 4 weeks of implantation, the value of the detachment force was higher in the case of 3D Ti implants functionalized with Li-C coatings in comparison to uncoated (control) ones. For the case of Li-P structures, this trend was similar, but there were particular instances when the results for coated and uncoated implants were very similar. The trend of higher detachment force was maintained for the case of 3D Ti implants functionalized with Li-P coatings as compared to the uncoated implants, however the effect seemed to be attenuated in respect to the Li-C coatings. After 9 weeks of implantation, it was observed that the detachment force was clearly superior (~1.3 times) for the 3D Ti implants functionalized with Li-C coatings as compared to the ones functionalized with Li-P (106.6 ± 2.9 N vs. 83.7 ± 2.1 N). A possible explanation for the higher values of the bone strength in the first case (Li-C coatings), could be connected to an increased crystallinity of the structures [57]. In this respect, it was shown that the Li-C coatings displayed a higher degree of crystallinity as compared to Li-P ones [33].

In the case of biological HA materials, as compared to synthetic HA, the osseoconduction rate is faster because the latter (i) has a higher solubility (which increases if its chemical structure contains carbonate groups), and (ii) contains Mg^{2+} and Na^{+} ions, which are two elements associated to bone remodeling [58]. This might represent one of the possible mechanisms through which the Li-C and Li-P coatings can improve the process of osseoconduction. One should therefore emphasize on the increased values of the detachment force obtained at 9 weeks, which demonstrate the capacity for long-term attachment to bone of our functionalized 3D metallic implants.

Usually, lithium is incorporated into CaPs in form of Li_2O [59,60]. In the current study, we selected to mix it with HA in form of Li_2CO_3 and Li_3PO_4. The reason was that in an initial parametric study (data not shown here), Li_2O mixed with natural origin HA displayed inferior results in vitro as compared to the Li_2CO_3 and Li_3PO_4 mixes. There were studies in the literature devoted to lithium relation to the human bone, some associating it with osteoporosis, while others signaling bone enhancement. In a study of Zamani et al. [61], it was shown that the treatment of patients with Li_2CO_3 preserved and even enhanced bone mass. Moreover, the results reported by Cohen et al. [62], demonstrated that Li_2CO_3 incorporated in drugs does not induce a risk of osteoporosis. One should note that, Li-P was tested in the current in vivo preliminary study because there are no reports in the dedicated literature related to its action on the bone. The results of the current work demonstrated that it possesses bioactive characteristics in conjunction with biological origin HA, however it was inferior to its Li-C counterpart.

All in all, the results of this preliminary in vivo assessment of the pulsed laser deposited BioHA doped with Li-C and Li-P coatings hold promise. Further and more insightful documentation on both the ingrowth characteristics' influence on the mechanical stability over various implantation time periods, and biomolecular analyses (i.e., related to gene expression of osteoblasts in contact with bone substitutes and/or HA [63]), will be considered. In addition, detailed in vivo studies, in which simple (undoped) BHA coatings will be used as controls, will be imagined not only to assess the lithium effectiveness in in vivo experiments, but to demonstrate their superiority also over the commercial Ti implants. All these investigations are necessary and should stand as the subject of a dedicated study which should start, however, from the results of this preliminary work.

5. Conclusions

Pulsed laser deposition was used for the synthesis of biological-derived hydroxyapatite doped with lithium carbonate (Li-C) and phosphate (Li-P) coatings. The Li-C and Li-P structures were investigated in vivo, as coatings onto 3D metallic implants which were inserted in rabbits' femoral condyles, for 4 and 9 weeks, respectively. The bone density measurements of the functionalized Ti implants, performed either at 4 or 9 weeks, showed superior values in comparison to simple (control) Ti ones. The inferred detachment force values of the functionalized Ti implants were ~2 times higher than those registered for the corresponding control ones. When referring to longer implantation time periods (i.e., 9 weeks), the extraction test results indicated improved bonding strength values (~5 times higher) of the functionalized Ti implants as compared to the same structures, but corresponding to an implantation time period of 4 weeks. Therefore, the mechanical testing is indicated as a promising tool to investigate the early phase of 3D Ti implants attachment to bone.

The demonstrated improvement of in vivo mechanical characteristics of the Li-C and Li-P synthesized coatings (as compared to control, commercial Ti implants), along with the fact that the materials used in this study were fabricated from natural, renewable sources, could stand for a pioneering contribution to the progress of medical devices. These important aspects should be taken into consideration when advancing this type of implant modification as an adequate choice to bare, non-deposited Ti implants for the manufacturing of a new generation of orthopedic implants, which will integrate better and much faster into the living body, corroborated with a substantially improved life-time.

6. Patents

A patent application entitled "Additive manufacturing of fixing devices for metallic implants by laser melting deposition method", by Chioibasu Diana, Mihai Sabin-Andrei, Duta Liviu-Marian, Popescu C. Andrei, containing information related to the procedure and devices used for fixing the 3D metallic implants both during the PLD experiments and during the post deposition thermal treatments, was submitted for evaluation to State Office for Inventions and Trademarks (OSIM), A00214/22.04.2020.

Author Contributions: Conceptualization, L.D.; methodology, L.D., O.A.Z., G.P.-P., D.C., and F.N.O.; software, L.D., J.N., R.P.M., O.A.Z., G.P.-P., D.C., and A.C.P.; validation, L.D., J.N., R.P.M. and A.C.P.; formal analysis, L.D.; investigation, J.N., R.P.M., and O.A.Z.; resources, L.D., F.N.O., and A.C.P.; data curation, L.D. and J.N.; writing—original draft preparation, L.D.; writing—review and editing, L.D., J.N., R.M., O.A.Z., G.P.-P, D.C., F.N.O., and A.C.P.; visualization, L.D., J.N., and A.C.P.; supervision, L.D.; project administration, L.D. and F.N.O.; funding acquisition, L.D. and A.C.P. All authors have read and agreed to the published version of the manuscript.

Funding: L.D. acknowledges the support from a grant of Ministry of Research and Innovation, CNCS-UEFISCDI, No. PN-III-P1-1.1-PD-2016-1568 (PD 6/2018), within PNCDI III. A.C.P. and D.C. have received funding from the PN-III-P1-1.1-TE-2016-2015 (TE136/2018) Project. This work was also supported by the Romanian Ministry of Education and Research, under Romanian National Nucleus Program LAPLAS VI—Contract 16N/2019.

Acknowledgments: All authors acknowledge George E. Stan for his useful comments and kind help.

Conflicts of Interest: The authors declare no conflict of interest.

References

1. Popa, A.C.; Stan, G.E.; Besleaga, C.; Ion, L.; Maraloiu, V.A.; Tulyaganov, D.U.; Ferreira, J.M.F. Submicrometer Hollow Bioglass Cones Deposited by Radio Frequency Magnetron Sputtering: Formation Mechanism, Properties, and Prospective Biomedical Applications. *ACS Appl. Mater. Interfaces* **2016**, *8*, 4357–4367. [CrossRef] [PubMed]
2. Stan, G.E.; Popescu, A.C.; Mihailescu, I.N.; Marcov, D.A.; Mustata, R.C.; Sima, L.E.; Petrescu, S.M.; Ianculescu, A.; Trusca, R.; Morosanu, C.O. On the bioactivity of adherent bioglass thin films synthesized by magnetron sputtering techniques. *Thin Solid Films* **2010**, *518*, 5955–5964. [CrossRef]
3. Šupová, M. Substituted hydroxyapatites for biomedical applications: A review. *Ceram. Int.* **2015**, *41*, 9203–9231. [CrossRef]
4. Graziani, G.; Boi, M.; Bianchi, M. A Review on Ionic Substitutions in Hydroxyapatite Thin Films: Towards Complete Biomimetism. *Coatings* **2018**, *8*, 269. [CrossRef]
5. Oladele, I.O.; Agbabiaka, O.; Olasunkanmi, O.G.; Balogun, A.O.; Popoola, M.O. Non-synthetic sources for the development of hydroxyapatite. *J. Appl. Biotechnol. Bioeng.* **2018**, *5*, 92–99. [CrossRef]

6. Akram, M.; Ahmed, R.; Shakir, I.; Ibrahim, W.A.W.; Hussain, R. Extracting hydroxyapatite and its precursors from natural resources. *J. Mater. Sci.* **2014**, *49*, 1461–1475. [CrossRef]
7. Tite, T.; Popa, A.C.; Balescu, L.M.; Bogdan, I.M.; Pasuk, I.; Ferreira, J.M.F.; Stan, G.E. Cationic substitutions in hydroxyapatite: Current status of the derived biofunctional effects and their in vitro interrogation methods. *Materials* **2018**, *11*, 2081. [CrossRef]
8. Ballini, A.; Mastrangelo, F.; Gastaldi, G.; Tettamanti, L.; Bukvic, N.; Cantore, S.; Cocco, T.; Saini, R.; Desiate, A.; Gherlone, E.; et al. Osteogenic differentiation and gene expression of dental pulp stem cells under low-level laser irradiation: A good promise for tissue engineering. *J. Biol. Regul. Homeost. Agents* **2015**, *29*, 813–822. [PubMed]
9. Bose, S.; Tarafder, S.; Bandyopadhyay, A. Hydroxyapatite coatings for metallic implants. In *Hydroxyapatite (Hap) for Biomedical Applications*; Mucalo, M., Ed.; Woodhead Publishing Series in Biomaterials; Elsevier: Amsterdam, The Netherlands, 2015; pp. 143–157. [CrossRef]
10. Barrere, F.; van der Valk, C.M.; Meijer, G.; Dalmeijer, R.A.; Groot, K.; Layrolle, P. Osteointegration of biomimetic apatite coating applied onto dense and porous metal implants in femurs of goats. *J. Biomed. Mater. Res. B Appl. Biomater.* **2003**, *67*, 655–665. [CrossRef]
11. Surmenev, R.A.; Surmeneva, M.A. A critical review of decades of research on calcium phosphate–based coatings: How far are we from their widespread clinical application? *Curr. Opin. Biomed. Eng.* **2019**, *10*, 35–44. [CrossRef]
12. Dorozhkin, S.V. Biphasic, triphasic, and multiphasic calcium orthophosphates. In *Advanced Ceramic Materials*; Tiwari, A., Gerhardt, R.A., Szutkowska, M., Eds.; Wiley, Scrivener Publishing: Austin, TX, USA, 2016; pp. 33–95. [CrossRef]
13. Crespi, R.; Capparè, P.; Gherlone, E. Comparison of magnesium-enriched hydroxyapatite and porcine bone in human extraction socket healing: A histologic and histomorphometric evaluation. *Int. J. Oral Maxillofac. Implant.* **2011**, *26*, 1057–1062. [PubMed]
14. Crespi, R.; Capparè, P.; Gherlone, E. Magnesium-enriched hydroxyapatite compared to calcium sulfate in the healing of human extraction sockets: Radiographic and histomorphometric evaluation at 3 months. *J. Periodontol.* **2009**, *80*, 210–218. [CrossRef] [PubMed]
15. Gherlone, E.F.; Capparé, P.; Tecco, S.; Polizzi, E.; Pantaleo, G.; Gastaldi, G.; Grusovin, M.G. A Prospective Longitudinal Study on Implant Prosthetic Rehabilitation in Controlled HIV-Positive Patients with 1-Year Follow-Up: The Role of CD4+ Level, Smoking Habits, and Oral Hygiene. *Clin. Implant Dent. Relat. Res.* **2016**, *18*, 955–964. [CrossRef] [PubMed]
16. Pesce, V.; Speciale, D.; Sammarco, G.; Patella, S.; Spinarelli, A.; Patella, V. Surgical approach to bone healing in osteoporosis. *Clin. Cases Miner. Bone Metab.* **2009**, *6*, 131–135. [PubMed]
17. León, B.; John, J. *Thin Calcium Phosphate Coatings for Medical Implants*; Springer-Verlag: New York, NY, USA, 2009; pp. 1–328. [CrossRef]
18. Available online: https://www.eos.info/en/3d-printing-examples-applications/people-health/medical-3d-printing (accessed on 9 October 2020).
19. Jiaxiang, Z.; Anh, Q.V.; Xin, F.; Suresh, B.; Michael, A.R. Pharmaceutical Additive Manufacturing: A Novel Tool for Complex and Personalized Drug Delivery Systems. *AAPS PharmSciTech* **2018**, *19*, 3388–3402. [CrossRef]
20. Sima, L.E.; Stan, G.E.; Morosanu, C.O.; Melinescu, A.; Ianculescu, A.; Melinte, R.; Neamtu, J.; Petrescu, S.M. Differentiation of mesenchymal stem cells onto highly adherent radio frequency-sputtered carbonated hydroxylapatite thin films. *J. Biomed. Mater. Res. A* **2010**, *95*, 1203–1214. [CrossRef]
21. Visan, A.; Grossin, D.; Stefan, N.; Duta, L.; Miroiu, F.M.; Stan, G.E.; Sopronyi, M.; Luculescu, C.; Freche, M.; Marsan, O.; et al. Biomimetic nanocrystalline apatite coatings synthesized by Matrix Assisted Pulsed Laser Evaporation for medical applications. *Mater. Sci. Eng. B Adv.* **2014**, *181*, 56–63. [CrossRef]
22. Graziani, G.; Bianchi, M.; Sassoni, E.; Russo, A.; Marcacci, M. Ion-substituted calcium phosphate coatings deposited by plasma-assisted techniques: A review. *Mater. Sci. Eng. C* **2017**, *74*, 219–229. [CrossRef]
23. Vranceanu, D.M.; Ionescu, I.C.; Ungureanu, E.; Cojocaru, M.O.; Vladescu, A.; Cotrut, C.M. Magnesium Doped Hydroxyapatite-Based Coatings Obtained by Pulsed Galvanostatic Electrochemical Deposition with Adjustable Electrochemical Behavior. *Coatings* **2020**, *10*, 727. [CrossRef]

24. Stewart, C.; Akhavan, B.; Wise, S.G.; Bilek, M.M.M. A review of biomimetic surface functionalization for bone integrating orthopedic implants: Mechanisms, current approaches, and future directions. *Prog. Mater. Sci.* **2019**, *106*, 100588. [CrossRef]
25. Asri, R.I.; Harun, W.S.; Hassan, M.A.; Ghani, S.A.; Buyong, Z. A review of hydroxyapatite-based coating techniques: Sol-gel and electrochemical depositions on biocompatible metals. *J. Mech. Behav. Biomed. Mater.* **2016**, *57*, 95–108. [CrossRef] [PubMed]
26. Chen, N.; Kim, D.H.; Kovacik, P.; Sojoudi, H.; Wang, M.; Gleason, K.K. Polymer thin films and surface modification by chemical vapor deposition: Recent progress. *Annu. Rev. Chem. Biomol. Eng.* **2016**, *7*, 373–393. [CrossRef]
27. Kumar, R.; Kumar, G.; Umar, A. Pulse laser deposited nanostructured ZnO thin films: A review. *J. Nanosci. Nanotechnol.* **2014**, *14*, 1911–1930. [CrossRef] [PubMed]
28. Chen, L.; Komasa, S.; Hashimoto, Y.; Hontsu, S.; Okazaki, J. In Vitro and In Vivo Osteogenic Activity of Titanium Implants Coated by Pulsed Laser Deposition with a Thin Film of Fluoridated Hydroxyapatite. *Int. J. Mol. Sci.* **2018**, *19*, 1127. [CrossRef] [PubMed]
29. Duta, L.; Popescu, A.C. Current Status on Pulsed Laser Deposition of Coatings from Animal-Origin Calcium Phosphate Sources. *Coatings* **2019**, *9*, 335. [CrossRef]
30. Bao, Q.; Chen, C.; Wang, D.; Ji, Q.; Lei, T. Pulsed laser deposition and its current research status in preparing hydroxyapatite thin films. *Appl. Surf. Sci.* **2005**, *252*, 1538–1544. [CrossRef]
31. Kuzanyan, A.S.; Kuzanyan, A.A. Pulsed Laser Deposition of Large-Area Thin Films and Coatings. In *Applications of Laser Ablation-Thin Film Deposition, Nanomaterial Synthesis and Surface Modification*; Yang, D., Ed.; InTech: Rijeka, Croatia, 2016. [CrossRef]
32. Greer, J.A. History and current status of commercial pulsed laser deposition equipment. *J. Phys. D Appl. Phys.* **2014**, *47*, 034005. [CrossRef]
33. Popescu, A.C.; Florian, P.E.; Stan, G.E.; Popescu-Pelin, G.; Zgura, I.; Enculescu, M.; Oktar, F.N.; Trusca, R.; Sima, L.E.; Roseanu, A.; et al. Physical-chemical characterization and biological assessment of simple and lithium-doped biological-derived hydroxyapatite thin films for a new generation of metallic implants. *Appl. Surf. Sci.* **2018**, *439*, 724–735. [CrossRef]
34. Florian, P.E.; Duta, L.; Grumezescu, V.; Popescu-Pelin, G.; Popescu, A.C.; Oktar, F.N.; Evans, R.W.; Constantinescu, A.R. Lithium-Doped Biological-Derived Hydroxyapatite Coatings Sustain In Vitro Differentiation of Human Primary Mesenchymal Stem Cells to Osteoblasts. *Coatings* **2019**, *9*, 781. [CrossRef]
35. Duta, L.; Chifiriuc, M.C.; Popescu-Pelin, G.; Bleotu, C.; Gradisteanu, G.P.; Anastasescu, M.; Achim, A.; Popescu, A. Pulsed Laser Deposited Biocompatible Lithium-Doped Hydroxyapatite Coatings with Antimicrobial Activity. *Coatings* **2019**, *9*, 54. [CrossRef]
36. Duta, L.; Oktar, F.N.; Stan, G.E.; Popescu-Pelin, G.; Serban, N.; Luculescu, C.; Mihailescu, I.N. Novel doped hydroxyapatite thin films obtained by pulsed laser deposition. *Appl. Surf. Sci.* **2013**, *265*, 41–49. [CrossRef]
37. Duta, L.; Mihailescu, N.; Popescu, A.C.; Luculescu, C.R.; Mihailescu, I.N.; Cetin, G.; Gunduz, O.; Oktar, F.N.; Popa, A.C.; Kuncser, A.; et al. Comparative physical, chemical and biological assessment of simple and titanium-doped ovine dentine-derived hydroxyapatite coatings fabricated by pulsed laser deposition. *App Surf. Sci.* **2017**, *413*, 129–139. [CrossRef]
38. Eisenhart, S. EU Regulation 722. In *New EU Animal Tissue Regulations in Effect for Some Medical Devices*; Emergo: Hong Kong, China, 2013; Available online: https://www.emergobyul.com/blog/2013/09/new-euanimaltissue-regulations-effect-some-medical-devices (accessed on 11 September 2020).
39. ISO 22442-1. *Medical Devices Utilizing Animal Tissues and Their Derivatives—Part 1: Application of Risk Management*; International Organization for Standardization: Berlin, Germany, 2015.
40. Chioibasu, D.; Mihai, S.-A.; Duta, L.; Popescu, A.C. Additive Manufacturing of Fixing Devices for Metallic Implants by "Laser Melting Deposition" Method. status (unpublished Patent application, A00214, submitted to State Office for Inventions and Trademarks, OSIM, 22 April 2020).
41. Lu, Y.P.; Chen, Y.M.; Li, S.T.; Wang, J.H. Surface nanocrystallization of hydroxyapatite coating. *Acta Biomater.* **2008**, *4*, 1865–1872. [CrossRef] [PubMed]
42. Neyt, J.G.; Buckwalter, J.A.; Carroll, N.C. Use of animal models in musculoskeletal research. *Iowa Orthop. J.* **1998**, *18*, 118–123.

43. Gilsanz, V.; Roe, T.F.; Gibbens, D.T.; Schulz, E.E.; Carlson, M.E.; Gonzalez, O.; Boechat, M.I. Effect of sex steroids on peak bone density of growing rabbits. *Am. J. Physiol.* **1988**, *255*, E416–E421. [CrossRef]
44. Salou, L.; Hoornaert, A.; Louarn, G.; Layrolle, P. Enhanced osseointegration of titanium implants with nanostructured surfaces: An experimental study in rabbits. *Acta Biomater.* **2015**, *11*, 494–502. [CrossRef] [PubMed]
45. Yan, J.; Sun, J.-F.; Chu, P.K.; Han, Y.; Zhang, Y.-M. Bone integration capability of a series of strontium-containing hydroxyapatite coatings formed by micro-arc oxidation. *J. Biomed. Mater. Res. A* **2013**, *101*, 2465–2480. [CrossRef]
46. Eason, R. *Pulsed Laser Deposition of Thin FILMS: Applications-Led Growth of Functional Materials*, 1st ed.; Wiley & Sons Interscience: Hoboken, NY, USA, 2007; pp. 1–705.
47. Pearce, A.I.; Richards, R.G.; Milz, S.; Schneider, E.; Pearce, S.G. Animal models for implant biomaterial research in bone: A review. *Eur. Cells Mater.* **2007**, *13*, 1–10. [CrossRef] [PubMed]
48. Goller, G.; Oktar, F.N.; Agathopoulos, S.; Tulyaganov, D.U.; Ferreira, J.M.F.; Kayali, E.S.; Peker, I. Effect of sintering temperature on mechanical and microstructural properties of bovine hydroxyapatite (BHA). *J. Sol-Gel Sci. Techn.* **2006**, *37*, 111–115. [CrossRef]
49. Sun, R.-X.; Lv, Y.; Niu, Y.-R.; Zhao, X.-H.; Cao, D.-S.; Tang, J.; Sun, X.-C.; Chen, K.-Z. Physicochemical and biological properties of bovine-derived porous hydroxyapatite/collagen composite and its hydroxyapatite powders. *Ceram. Int.* **2017**, *43*, 16792–16798. [CrossRef]
50. Rahavi, S.S.; Ghaderi, O.; Monshi, A.; Fathi, M.H. A comparative study on physicochemical properties of hydroxyapatite powders derived from natural and synthetic sources. *Russ. J. Non-Ferrous Metals* **2017**, *58*, 276–286. [CrossRef]
51. Ofudje, E.A.; Rajendran, A.; Adeogun, A.I.; Idowu, M.A.; Kareem, S.O.; Pattanayak, D.K. Synthesis of organic derived hydroxyapatite scaffold from pig bone waste for tissue engineering applications. *Adv. Powder Technol.* **2017**, *29*, 1–8. [CrossRef]
52. Su, Y.; Cockerill, I.; Zheng, Y.; Tang, L.; Qin, Y.-X.; Zhu, D. Biofunctionalization of metallic implants by calcium phosphate coatings. *Bioact. Mater.* **2019**, *4*, 196–206. [CrossRef]
53. Sul, Y.-T.; Johansson, C.; Albrektsson, T. A novel in vivo method for quantifying the interfacial biochemical bond strength of bone implants. *J. R Soc. Interface* **2009**, *7*, 81–90. [CrossRef] [PubMed]
54. Shannon, F.J.; Cottrell, J.M.; Deng, X.-H.; Crowder, K.N.; Doty, S.B.; Avaltroni, M.J.; Warren, R.F.; Wright, T.M.; Schwartz, J. A novel surface treatment for porous metallic implants that improves the rate of bony ongrowth. *J. Biomed. Mater. Res. A* **2008**, *86*, 857–864. [CrossRef]
55. Schumacher, T.C.; Tushtev, K.; Wagner, U.; Becker, C.; Große Holthaus, M.; Hein, S.B.; Haack, J.; Engelhardt, C.H.M.; Khassawna, T.E.; Rezwan, K. A novel, hydroxyapatite-based screw-like device for anterior cruciate ligament (ACL) reconstructions. *Knee* **2017**, *24*, 933–939. [CrossRef]
56. Aparicioa, C.; Padrósb, A.; Gil, F.-J. In vivo evaluation of micro-rough and bioactive titanium dental implants using histometry and pull-out tests. *J. Mech. Behav. Biomed.* **2011**, *4*, 1672–1682. [CrossRef]
57. Kumar, R.R.; Maruno, S. Functionally graded coatings of HA–G–Ti composites and their in vivo studies. *Mater. Sci. Eng. A Struct.* **2002**, *334*, 156–162. [CrossRef]
58. Hayami, T.; Hontsu, S.; Higuchi, Y.; Nishikawa, H.; Kusunoki, M. Osteoconduction of a stoichiometric and bovine hydroxyapatite bilayer-coated implant, hydroxyapatite bilayer-coated implant. *Clin. Oral Implant. Res.* **2011**, *22*, 774–776. [CrossRef]
59. Zhang, K.; Alaohali, A.; Sawangboon, N.; Sharpe, P.T.; Brauer, D.S.; Gentleman, E. A comparison of lithium-substituted phosphate and borate bioactive glasses for mineralised tissue repair. *Dent. Mater.* **2019**, *35*, 919–927. [CrossRef]
60. Khan, P.K.; Mahato, A.; Kundu, B.; Nandi, S.K.; Mukherjee, P.; Datta, S.; Sarkar, S.; Mukherjee, J.; Nath, S.; Balla, V.K.; et al. Influence of single and binary doping of strontium and lithium on in vivo biological properties of bioactive glass scaffolds. *Sci. Rep. UK* **2016**, *6*, 32964. [CrossRef]
61. Zamani, A.; Omrani, G.R.; Nasab, M.M. Lithium's effect on bone mineral density. *Bone* **2009**, *44*, 331–334. [CrossRef]

62. Cohen, O.; Rais, T.; Lepkifker, E.; Vered, I. Lithium Carbonate Therapy is not a Risk Factor for Osteoporosis. *Horm. Metab. Res.* **1998**, *30*, 594–597. [CrossRef] [PubMed]
63. Crespi, R.; Capparé, P.; Romanos, G.E.; Mariani, E.; Benasciutti, E.; Gherlone, E. Corticocancellous porcine bone in the healing of human extraction sockets: Combining histomorphometry with osteoblast gene expression profiles in vivo. *Int. J. Oral Maxillofac. Implant.* **2011**, *26*, 866–872. [PubMed]

Publisher's Note: MDPI stays neutral with regard to jurisdictional claims in published maps and institutional affiliations.

Article

Improvement of CoCr Alloy Characteristics by Ti-Based Carbonitride Coatings Used in Orthopedic Applications

Mihaela Dinu [1], Iulian Pana [1], Petronela Scripca [1], Ioan Gabriel Sandu [2], Catalin Vitelaru [1] and Alina Vladescu [1,3,*]

[1] Department for Advanced Surface Processing and Analysis by Vacuum Technologies, National Institute of Research and Development for Optoelectronics-INOE 2000, 409 Atomistilor St., 077125 Magurele, Romania; mihaela.dinu@inoe.ro (M.D.); iulian.pana@inoe.ro (I.P.); petronela.scripca@inoe.ro (P.S.); catalin.vitelaru@inoe.ro (C.V.)

[2] Faculty of Material Sciences and Engineering, Gheorghe Asachi Technical University of Iasi, 41 Prof.dr.doc. D. Mangeron St., 700050 Iasi, Romania; gisandu@yahoo.com

[3] Physical Materials Science and Composite Materials Centre, National Research Tomsk Polytechnic University, Lenin Avenue 43, 634050 Tomsk, Russia

* Correspondence: alinava@inoe.ro; Tel.: +4-21-457-57-59

Received: 15 April 2020; Accepted: 19 May 2020; Published: 22 May 2020

Abstract: The response of the human body to implanted biomaterials involves several complex reactions. The potential success of implantation depends on the knowledge of the interaction between the biomaterials and the corrosive environment prior to the implantation. Thus, in the present study, the in vitro corrosion behavior of biocompatible carbonitride-based coatings are discussed, based on microstructure, mechanical properties, roughness and morphology. TiCN and TiSiCN coatings were prepared by the cathodic arc deposition method and were analyzed as a possible solution for load bearing implants. It was found that both coatings have an almost stoichiometric structure, being solid solutions, which consist of a mixture of TiC and TiN, with a face-centered cubic (FCC) structure. The crystallite size decreased with the addition of Si into the TiCN matrix: the crystallite size of TiCN was 16.4 nm, while TiSiCN was 14.6 nm. The addition of Si into TiCN resulted in smaller R_a roughness values, indicating a beneficial effect of Si. All investigated surfaces have positive skewness, being adequate for the load bearing implants, which work in a corrosive environment. The hardness of the TiCN coating was 36.6 ± 2.9 GPa and was significantly increased to 47.4 ± 1 GPa when small amounts of Si were added into the TiCN layer structure. A sharp increase in resistance to plastic deformation (H^3/E^2 ratio) from 0.63 to 1.1 was found after the addition of Si into the TiCN matrix. The most electropositive value of corrosion potential was found for the TiSiCN coating (−14 mV), as well as the smallest value of corrosion current density (49.6 nA cm^2), indicating good corrosion resistance in 90% DMEM + 10% FBS, at 37 ± 0.5 °C.

Keywords: titanium-based carbonitrides; coating; corrosion resistance; X-ray diffraction; nanoindentation; cathodic arc deposition

1. Introduction

In medical applications, especially for hip and knee implants, a CoCr alloy is mostly used, due to good mechanical, anticorrosive and tribological characteristics [1–3]. CoCr alloys have also been used as screws in trauma plating systems. In this case, a low osseointegration, compared to Ti alloys, could facilitate an easier removal after the healing of the bone fracture. Due to its higher strength, a CoCr alloy was also used for idiopathic scoliosis applications, where the results proved to be better in the case of the CoCr alloy compared to stainless steel (SS) and a Ti alloy [4]. Another application was the use of a CoCr alloy in implants used for the correction of spine deformities, due to the high rigidity of the CoCr rod compared to SS- and Ti-based ones [5,6]. CoCr was also found to be well adapted in dentistry for its good castability, especially the wrought alloys. Guide wires, clips, orthodontic arch wires and catheters are among the main applications in this field [7,8]. Thus, the decreased corrosion resistance of SS and the low wear resistance of Ti alloys make it difficult to replace CoCr alloys in a wide range of applications.

Despite these good properties, in clinical practice the implants made of a CoCr alloy exhibited a high rate of failure due to various complications after implantation: (i) high toxicity as a result of the migration of toxic metal ions; (ii) a high amount of wear debris surrounding the peri-implant tissues and body organs, mainly due to the low wear resistance in biological fluids; (iii) low bioactivity abilities; (iv) a non-hydrophilic surface [1,2]. During the friction process, wear debris of CoCr alloys were generated in different size and shapes [9] and migrated to the periprosthetic tissues, leading to a failure of the implants. The wear of hip or knee joints is a complex process that involves many factors, such as the material and geometry of the implant, synovial fluid properties (various protein levels) and the patients' lifestyles and body weight. Thus, the wear particles larger than 0.5–10 μm (round to oval to irregular shapes) play a dramatic role as third-body wear, leading to an intense wear process [9–11]. The main problem with these debris is their high size, as cells (i.e., macrophages, fibroblasts, giant cells, neutrophils, lymphocytes, osteoclasts) will interact with these debris, leading to a chronic inflammatory response [9,12–14]. Another problem can be the release of Co and Cr ions into the synovial fluid and their correlated increased concentration in the blood. Although considered essential elements for the body, their increase in concentration can be detrimental for certain functions [15]. Thus, their long-term exposure can lead to cellular effects in the adjacent tissue and even to necrosis [16]. Nevertheless, in order to eliminate the mentioned disadvantages found in CoCr alloy medical implants, several modifications of alloy surfaces were carried out over recent decades by: (i) coating, using various types of techniques (PVD, CVD, ion implantation, plasma spray); (ii) surface structuring (laser processing, sand blasting, acid etching, anodization); (iii) micro arc oxidation; (iv) electrochemical oxidation [1,17]. The PVD method chosen for this study, namely the cathodic arc method, combines both a high degree of ionization of the ejected particles and a high efficiency of the evaporation process. Even though the initial energies are about 20 eV for light elements and around 200 eV for heavy elements, the final ion velocities (in the range of $1–2 \times 10^4$ m/s) were found to be independent of the cathode material and ion charge state, due to electron–ion coupling [18,19]. Thus, with an enhanced atom mobility and surface diffusion, due to the higher energy of the ions, the obtained materials have favorable conditions in order to obtain different coating properties [20].

The present study aimed to analyze TiSiCN as a possible coating solution to improve the corrosion and wear resistance of CoCr alloys used for orthopedic implants. For comparison, TiCN and uncoated CoCr were used as control groups. The TiCN coating was selected as a reference because it has good mechanical properties and good corrosion resistance, and an acceptable wear resistance in dry environments [21–23]. By the addition of Si into TiCN, it was expected to significantly reduce the friction and wear process, as well as to improve the corrosion resistance of the CoCr alloy. It was reported that the addition of Si governs grain refinement and Si-containing coatings present superior friction and wear performance [24]. Furthermore, it was demonstrated that the addition of Si to various materials with biological applications enhances the proliferation and differentiation of human osteoblasts, accelerating the osseointegration process [25]. Additionally, Si–N thin films proved to have remarkable properties, which included high thermal stability and chemical inertness, in addition to those already mentioned [26]. A survey of the literature shows that TiSiCN coatings have a superior tribological performance, but tests were performed mainly in conditions used in industrial applications, such as cutting tools and the automotive industry. Their main advantages are low friction, high wear resistance, good mechanical properties such as toughness and high resistance on plastic deformation [27–32]. For medical applications, however, they have not yet been tested. Nevertheless, an alternative solution was the addition of Zr and Cr to the Si–N and Si–C–N matrix for severe wear and corrosive applications [33].

In the present study, the coatings were obtained by the cathodic arc evaporation method on the CoCr substrates under a mixture of CH_4 and N_2 gases. The investigation included the examination of elemental and phase composition, texture, structure, morphology and mechanical properties (Young modulus, hardness, roughness, stress). Special attention was devoted to the corrosion resistance performed in a 90% DMEM + 10% FBS solution, at 37 ± 0.5 °C. In order to understand the damaging effect of the corrosion test, the morphology and roughness after corrosion were evaluated. Electrochemical impedance spectroscopy (EIS) was also performed in order to investigate the behavior of the proposed systems. Thus, this method gave an insight into the electrochemical processes which occurred at the material–electrolyte interfaces. The research was conducted in order to find new and improved structures as a better solution to optimize the safety and efficacy of biomaterials.

2. Materials and Methods

2.1. Coatings Preparation

Both coatings were prepared by the cathodic arc deposition process. For TiCN, one Ti cathode (99.99% purity) was used, while for TiSiCN, a Ti+Si cathode (85 at.% Ti and 15 at.% Si; 99.9% purity) was used. The cathodes were supplied from Cathay Advanced Materials Limited, Guangdong, China. The coatings were prepared using both CH_4 and N_2 as reactive gases (99.999% purity, Linde). The position of each cathode inside the deposition chamber was the same as those described in [33]. In order to guarantee the uniform thickness of the coatings, the substrate holder was rotated by 15 rot/min. The CoCr substrates (ASTM F75 CoCr alloy) were cut into 12 mm discs and polished up to Ra roughness of 46.9 ± 5.9 nm. Each substrate was ultrasonically cleaned in trichloroethylene and flushed with dry nitrogen, and then introduced in the deposition chamber. To eliminate any impurity, the substrates were sputter etched with Ar^+ ions (1 keV) for 15 min. The deposition parameters are listed in Table 1, and were maintained constant during all deposition runs. The same negative substrate bias voltage was applied on both cathodes and the substrate temperature was around 320 °C. The thickness of the coatings was around 2.5 μm.

Table 1. Conditions for the developed coatings.

<table>
<tr><th>Deposition Parameters</th><th>TiCN</th><th>TiSiCN</th></tr>
<tr><td>Base pressure</td><td colspan="2">2×10^{-3} Pa</td></tr>
<tr><td>Working pressure</td><td>2×10^{-2} Pa</td><td>6×10^{-2} Pa</td></tr>
<tr><td>CH_4 mass flow rate</td><td colspan="2">80 sccm</td></tr>
<tr><td>N_2 mass flow</td><td colspan="2">120 sccm</td></tr>
<tr><td>Arc current on each cathode</td><td colspan="2">90 A</td></tr>
<tr><td>Substrate bias voltage</td><td colspan="2">−150 V</td></tr>
<tr><td>Deposition duration</td><td colspan="2">40 min</td></tr>
</table>

2.2. Coatings Characterization

Energy dispersive spectrometry (EDS) was used for the analysis of the elemental composition (EDS, Quantax70, Bruker, Tokyo, Japan). The morphology was examined by scanning electron microscopy (SEM, Hitachi TM3030Plus, Tokyo, Japan). Phase composition was studied by the X-ray diffraction method (XRD, SmartLab diffractometer, Rigaku, Tokyo, Japan), using Cu Kα radiation, from 10° to 100° with a step size of 0.02°/min. For the thickness and surface roughness determination, a surface profilometer (Dektak 150, Bruker, Billerica, MA, USA) was used. Surface roughness was measured for each investigated sample on five line-scans, each on a distance of 4000 μm.

The mechanical properties of the coatings were determined using a Hysitron Premier TI nanoindentation unit equipped with a Berkovich indenter tip of 100 nm radius and a total included angle of 142.3°, respectively. Prior to any sample testing, the Z-axis calibration was performed in the air and the machine compliance was assured using a fused quartz standard calibration sample with known hardness (H = 9.25 GPa ± 10%) and elastic modulus (E = 69.6 GPa ± 10%). In order to perform the nanoindentation experiments, a 15 × 15 μm^2 area was previously scanned using the same Berkovich diamond tip at a normal force of 2 μN to investigate the surface roughness for the subsequent indents. Additionally, the indents were intentionally located at least 5 μm apart from each other, whilst an applied force of 10 mN was employed for every nanoindentation test. The force–displacement curves were recorded using a gradual force increase up to 10 mN in a 7-s time interval, followed by a 2-s dwell time at the maximum force of 10 mN and a gradual force decrease within the next 7 s, until the complete tip retraction from the coatings surface.

The electrochemical behavior of the investigated specimens was analyzed by potentiodynamic polarization and electrochemical impedance spectroscopy (EIS), using a VersaSTAT 3 Potentiostat/Galvanostat system. The measurements were performed in 90% DMEM + 10% FBS, at 37 ± 0.5 °C, using a typical three electrode setup with a Pt grid counter electrode (CE) and a saturated calomel (saturated KCl) (SCE) as the reference electrode (RE), while the working electrode consisted of uncoated CoCr and TiCN or TiSiCN coated CoCr substrates, respectively.

The open circuit potential (E_{OC}) was monitored for 1 h, starting after the sample's immersion. Linear polarization, Tafel and potentiodynamic curves were performed by applying a potential of −20 to 20 mV vs. E_{OC}, −50 to 250 mV vs. E_{OC} and −1 V vs. E_{OC} to 2 V vs. RE, respectively, with a scanning rate of 0.167 mV/s. For the linear polarization measurements, the testing conditions were selected based on preliminary results, in such a way that the applied potential gave a linear behavior. The selected value was used to accommodate all the investigated systems. The EIS measurements were performed over a range of frequencies (0.1 ÷ 10^4 Hz), by applying a sinusoidal signal of 10 mV RMS vs. E_{OC}.

3. Results

3.1. Elemental and Phase Composition

The elemental composition of the coatings obtained by the EDS method is presented in Table 2. Both coatings had an almost stoichiometric structure. The amount of Si was low but proved to be sufficient for the goal of this study. Both coatings exhibited a preferred growth orientation after (111) plan (Figure 1). The TiCN peaks were located between those of TiC and TiN, indicating that the TiCN is a solid solution and consisted of a mixture of TiC and TiN, with a face-centered cubic (FCC) structure (like NaCl). The same results were also reported by other authors regarding the TiCN coatings [21]. The identification of TiCN was performed according to JCPDS no. 042-1488 (red lines in Figure 1). It can be seen that our TiCN is shifted towards a lower Bragg angle compared with the TiCN standard, which is related to the formation of more TiC phase (Figure 1). This finding is in accordance with the results reported by Wang et al. [34] regarding the TiSiCN coatings. The preferred orientation of both coatings was also confirmed by the texture coefficient T(*hkl*), a strong texture intensity at (111) plane being found in the case of both coatings. Karlsson et al. reported that the plane with a preferred orientation parallel to the investigated surface has a texture coefficient value higher than 1 [21].

Table 2. Elemental composition measured by EDS, texture coefficient T(*hkl*), crystallite size ***d*** and strain ε determined by the Williamson–Hall plot method.

Coating	Elemental Composition (at.%)				(C+N)/Σ(Me+Si)	T(*hkl*)				d (nm)	ε
	Ti	Si	C	N		(111)	(200)	(220)	(311)		
TiCN	52.1	-	17.5	30.4	0.9	0.59	0.09	0.24	0.07	16.4	0.053
TiSiCN	48.4	3.4	20.1	28.1	0.9	0.73	0.06	0.17	0.05	14.6	0.012

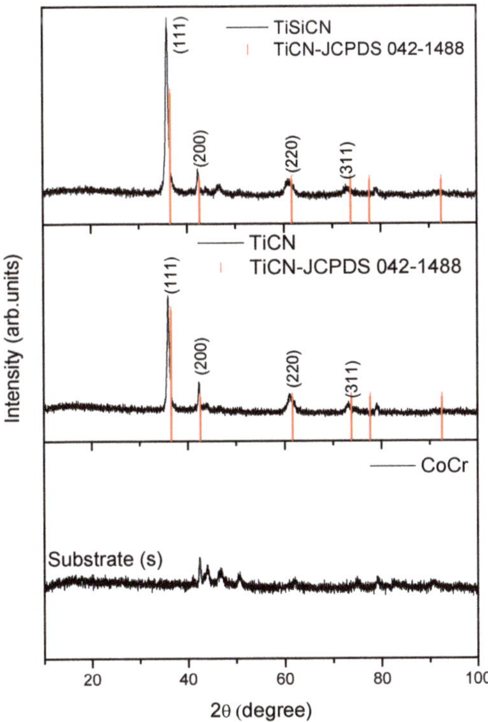

Figure 1. X-ray diffraction patterns of the investigated coatings (S-substrate) as well as of the TiCN standard (JCPDS 042-1488).

The crystallite size was determined by the Debye–Scherrer equation of the peak (111) and is presented in Table 2. The crystallite size decreased from 16.4 nm for TiCN coatings to 14.6 nm for TiSiCN. The possible explanation is the penetration of impinging Si ions into the lattice of the TiCN coating, and the decreased generation of defects, led to a decreased number of preferential nucleation sites causing reduced grain [35,36]. The strain ε was determined by the Williamson–Hall plot method and is presented in Table 2. In both cases, the strain was low, but after the addition of Si, the strain was significantly reduced. It is reasonable to believe that the observed difference in strain of the TiSiCN coatings is related to the amorphous phases in which Si can be found (Si, SiNx, SiCN) or the C=C phase at grain boundaries, which is detrimental for crystallite development. These effects will be goal of another paper, in which other complex analyses, such as TEM and XPS, will be carried out.

The lattice parameter obtained from 2θ values of the (111), (200), (220) and (311) of TiSiCN was 4.2771, while for TiCN it was 4.2725. According to JCPDS no. 042-1488, TiCN has a lattice parameter equal to 4.2644. When the diffraction peak is shifted towards lower Bragg angle, an increase of lattice parameter is observed. This result is in good agreement with those of Constantin et al. related to TiCN coatings [37]. Both coatings exceeded the size of the standard unit cell parameters. This finding can be a sign that the hardness can be low and, on the other hand, the toughness will be high [38].

Grieveson et al. stated that the TiN and TiC compounds do not combine perfectly; the mixture is far from the ideal Raoultian behavior [39,40]. The same conclusion was also considered by Levi et al., who revealed that the occupation of all sites by Ti, C and N is not random, it is a TiN-based structure with a replacement of the N site by C, forming an FCC or tetragonal structures on the low concentration of vacancies that might be present [41]. All these aspects had an important effect on the characteristics of the TiCN coatings.

3.2. Morphology and Roughness (Before Corrosion)

The surface morphologies of the uncoated CoCr substrate and the TiCN and TiSiCN coatings can be observed in Figure 2. The appearance of the coatings is characterized by a continuous coverage of the substrate with visible individual microparticles distributed on the surface. These droplets are considered defects and are characteristic of the energetic ejection of the particles during the arc deposition process [42,43].

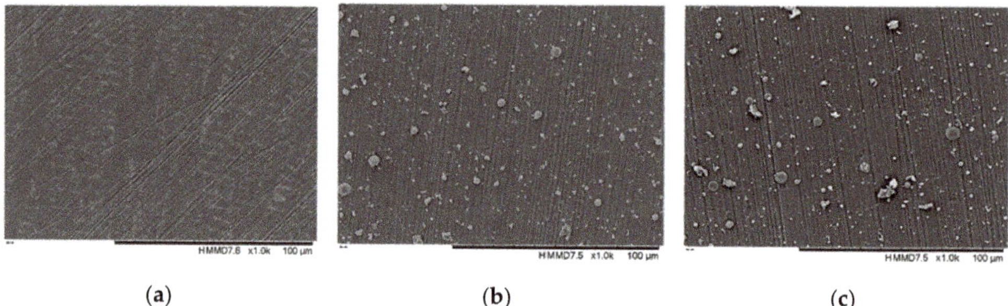

Figure 2. SEM images of the (**a**) uncoated CoCr substrate and (**b**) TiCN and (**c**) TiSiCN coatings.

3.3. Roughness of Surface

The main surface texture parameters of the investigated surfaces are: R_a: the arithmetic average of the roughness profile, R_q: the root mean square average of the roughness profile, S_k: skewness (according to the ISO 2517 standard [44]), and are presented in Table 3. The uncoated substates had an R_a roughness around 50 µm, as intended. The roughness of the coated surfaces was much higher and it was more evident in the TiCN coated surfaces. Similar values of both R_a and R_q were found for the TiCN and TiSiCN coatings.

Table 3. Main roughness parameters of the investigated specimens.

Sample	CoCr	TiCN	TiSiCN
R_a (nm)	46.9 ± 5.9	534.6 ± 111.4	499.7 ± 28.9
R_q (nm)	59.5 ± 7.0	746.8 ± 181.9	650.4 ± 30.7
S_k	0.2 ± 0.4	2.1 ± 0.3	1.6 ± 0.1

A negative value of S_k shows that the surface consists of many valleys, while a surface with a positive S_k contains mainly peaks and asperities. Therefore, a surface with positive S_k has a good tribological performance in lubrication conditions. Taking into account this observation, both investigated coatings exhibited a positive S_k, which is suitable for load bearing implants, which work in a corrosive environment. The uncoated substrate had an S_k close to 0, meaning that the surface was very flat. The TiCN surfaces had a high S_k, indicating many peaks and asperities, whereas Si-containing coatings had a smaller value, indicating a decrease in peaks and asperities. The dependence of corrosion resistance on these parameters will be discussed below.

3.4. Hardness and Elastic Modulus

Nanoindentation is a widely used technique, and it is a powerful tool to investigate the nanomechanical properties of materials, such as hardness and elasticity [45]. In order to overcome the substrate and roughness effects on nanoindentation tests, many studies have pointed out the importance of two basic rules. They refer to the maximum penetration depth, which should not exceed 1/10 of the layer thickness, and should be at least 20 times higher than the average roughness of the indented surfaces [46,47]. Another important limitation that needs to be considered for a reliable nanoindentation test is represented by the tip geometry and radius used, which limits the good testing contact depths, hence, in our case, was necessary to obtain contact depths higher than 40 nm [48].

In view of these testing conditions, the hardness and reduced modulus of the investigated sample coatings were determined according to the Oliver–Pharr method using similar force–displacement curves to the ones presented in Figure 3a. As it can be seen, these curves were obtained in both elastoplastic (CoCr substrate) and nearly elastic regimes (TiCN and TiSiCN coatings) [49], which were used to obtain the mechanical parameters of the samples. Note that all experimental force–displacement curves were qualitatively similar for the TiCN and TiSiCN coatings, providing indentation depth values smaller than 130 nm and exhibiting well-defined edges of the residual indent impressions (inset in Figure 3a). Moreover, the relatively low surface roughness in the investigated regions of the samples (~5 nm) was confirmed by performing several scans on 4 μm^2 areas using the same Berkovich tip. It has to be noted that this roughness corresponds to the flat areas between the microparticles found on the surface and it is therefore much smaller than the values reported for 4000 µm length surface scans. The resulting H = 7 ± 0.2 GPa and E = 97 ± 1.1 GPa values for the CoCr substrate are much smaller than the ones obtained for the investigated coatings. The hardness of the coatings significantly increased from 36.6 ± 2.9 GPa for the TiCN coating to 47.4 ± 1 GPa for TiSiCN, despite the small amounts of Si added into the TiCN layer structure (Figure 3b), while the effective modulus changed from 277 ± 8 GPa to 310 ± 2 GPa. The hardness enhancement originates from the grain size effect ($H \sim d^{-1/2}$, according to the Hall–Petch strengthening mechanism), since the TiSiCN coatings have smaller grain sizes. The hardness and modulus values were consistent with the previous results [50,51]. This implies that the condition H/E > 0.1 [52] was fulfilled for both coatings, as the calculated values for TiCN and TiSiCN were about 0.13 and 0.15, respectively. As the TiCN structure was modified, the sharp increase in the H^3/E^2 ratio (resistance to plastic deformation) from 0.63 to 1.1, testifies to the superior toughness of the TiSiCN layer [53], and it may be ascribed to the beneficial effect of the Si content. The H^3/E^2 parameter was considered to be a parameter sensitive to the tribological and corrosion properties of the materials [54].

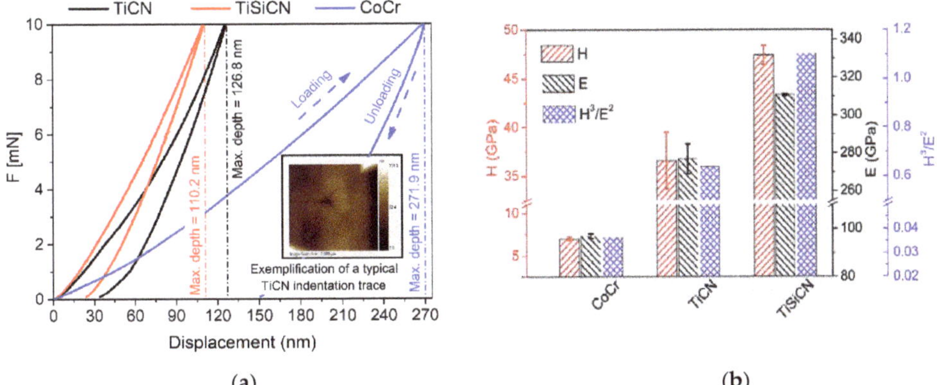

Figure 3. (**a**) Experimental force–displacement curves of the CoCr substrate, TiCN and TiSiCN coatings and (**b**) averaged hardness and reduced Young's modulus values.

3.5. *In Vitro Corrosion Investigations*

3.5.1. Potentiodynamic Polarization

The open circuit potential (E_{OC}) evolution after 1 h of immersion and the potentiodynamic curves of the investigated systems are presented in Figure 4. During the immersion, the E_{OC} of the TiCN and TiSiCN thin films showed a steady evolution, while the substrate slightly changed its value for half of the time, reaching a stable evolution at the end of the test. The TiSiCN coatings exhibited a positive

E_{OC} value (54 mV) compared with the TiCN coatings, indicating that the addition of Si had a positive effect on corrosion behavior.

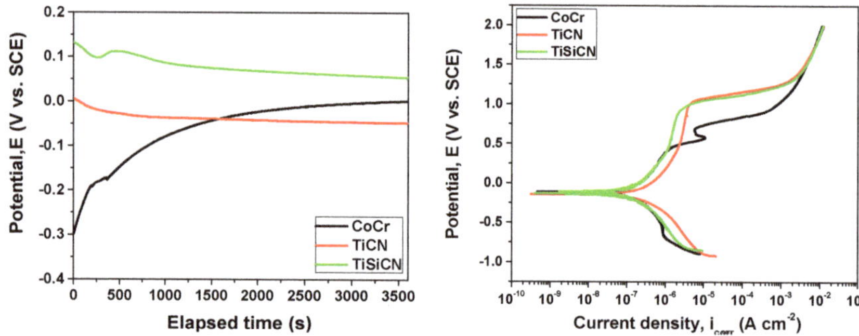

Figure 4. (**a**) Open circuit potential evolution in time and (**b**) potentiodynamic curves of the investigated systems.

The main electrochemical parameters of the investigated specimens calculated based on the Tafel and potentiodynamic curves are presented in Table 4 (all the presented potentials are relative to the SCE value). Polarization resistance (R_p) was determined from linear polarization measurements as the slope of the linear region of the ΔE–Δi curve near E_{corr}. Corrosion potential (E_{corr}), anodic (β_a) and cathodic (β_c) slopes were estimated from Tafel plots. The corrosion current density was also calculated based on one form of the Stern–Geary equation (Equation (1)), based on previously determined parameters.

$$\frac{1}{R_p} = \left(\frac{\Delta i}{\Delta E}\right)_{E_{corr}} = 2.3\left(\frac{\beta_a + |\beta_c|}{\beta_a\,|\beta_c|}\right) i_{corr} \quad (1)$$

considering the E_{corr} parameter, the most electropositive value was demonstrated by the TiSiCN coating ($E_{corr\ TiSiCN}$ = −14 mV), as well as the smallest value of corrosion current density ($i_{corr\ TiSiCN}$ = 49.6 nA cm^{-2}). Taking into account the polarization resistance, the highest value was observed in the case of the CoCr substrate, closely followed by the TiSiCN coating ($R_{p\ TiSiCN}$ = 425 kΩ·cm^2).

Table 4. The main corrosion parameters of the investigated specimens.

Substrate/ Coating	E_{oc} (mV)	R_p (kΩ cm^2)	E_{corr} (mV)	β_a (mV dec^{-1})	β_c (mV dec^{-1})	i_{corr} (nA cm^{-2})
CoCr	1	455	−36	516	447	198.2
TiCN	−48	216	−103	517	288	233.0
TiSiCN	54	425	−14	271	269	49.6

3.5.2. Electrochemical Impedance Spectroscopy (EIS)

It was observed that the electrochemical performance of the investigated specimens was different as a function of their composition. For comparison, Nyquist and Bode plots for the investigated specimens are presented in Figure 5.

The electrochemical parameters were obtained by fitting the data with an equivalent circuit, which took into consideration the phenomenon at the interface of each investigated system with the testing electrolyte (inset Figure 5). R_{el} represents the electrolyte resistance, CPE_{layer} represents the coating capacitance, R_{pore} represents the resistance associated with the current flow through the pores generated by the coatings' defects and CPE_{dl} is a double layer capacitance in parallel with a charge transfer resistance - R_{ct}. CPE was used instead of a capacitor due to the non-ideal character of the working electrode. The physical interpretation of a circuit that has a constant phase element (for a better quality fit) depends on the value of α. If the α parameter is 1, then the CPE can be modeled as a capacitor. Since after the fitting, the α parameter showed values less than 1 in both cases (i.e., the α layer and α_{dl}), a CPE was used. This can be due to possible deviations from the ideal dielectric behavior and it is usually related to the surface inhomogeneity [55]. According to Hirschorn et al. [56], these deviations arise either from different properties along the surface of an electrode (e.g., roughness), or properties normal for the surface (e.g., thickness).

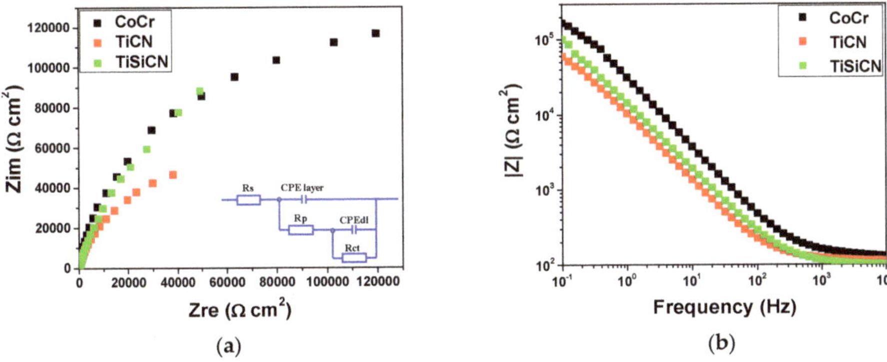

Figure 5. (**a**) Nyquist plot (electrical circuit used for the fitting procedure included) and (**b**) Bode plot of the investigated specimens.

The electrochemical parameters of the investigated systems are presented in Table 5. It can be noticed the low value of the χ^2 parameter, which is an indication of an excellent agreement between the experimental data and those simulated by the equivalent circuit.

Table 5. The fitting results of EIS curves for the investigated systems.

Substrate/ Coating	R_{el} ($\Omega\cdot cm^2$)	Q_{layer} ($\mu Fs^{(\alpha-1)}$ cm^{-2})	α Layer	R_{pore} ($k\Omega\cdot cm^2$)	Q_{dl} ($\mu Fs^{(\alpha-1)}$ cm^{-2})	α dl	R_{ct} ($k\Omega\cdot cm^2$)	χ^2
CoCr	127	1.94	0.96	88	3.66	0.91	243,830	13×10^{-3}
TiCN	113	16.19	0.87	216	2.19	0.99	112,000	3×10^{-3}
TiSiCN	101	12.38	0.87	316	1.21	0.96	262,750	5×10^{-3}

Taking into account the fitting results for the investigated systems, it can be noticed that the highest pore-associated resistance was obtained for TiSiCN, while the CoCr substrate showed low values. It was stated that CoCr alloys form a passive layer at the surface, which is mainly based on Cr(III), and smaller amounts of $Cr(OH)_3$, Co and Mo oxides [57]. The fitted values associated to the CoCr specimen showed that the formed layer is not as compact and protective as the TiCN and TiSiCN coatings.

The Q_{dl} parameter, which is representative of the substrate–electrolyte interface, indicated a better protection of the deposited/formed layer in the following order: TiSiCN > TiCN > CoCr. Thus, the best protection after immersion in 90 % DMEM + 10 % FBS was observed for the TiSiCN coatings, with the best capacitive character, indicated by the low value of Q_{layer} and the highest value of R_{ct}.

Considering α_{layer}, it can be observed that for CoCr, the CPE used for fitting the obtained data were the closest to a capacitor, since in this case, $\alpha_{layer} = 0.96$. This could be due to the low roughness measured before the corrosion as compared to the other investigated specimens (Table 3). Similar α layer values were obtained for the coatings and the time-constant dispersion is ascribed to the similar values of roughness. Going deeper, at the interface between the coating and the substrate, another double layer is formed. Q_{dl} and α_{dl} can give an indication of the compactness of the deposited/formed layer and the electrolyte ingress through the defects, which can create pathways for the electrolyte to reach the substrate [58]. It can be observed that even though α_{dl} was higher for TiCN, showing an almost defect-free structure, TiSiCN was the one showing better values of Q_{dl} and R_{ct}.

3.6. Morphology and Roughness after Corrosion

SEM images after the corrosion tests are presented in Figure 6. It is worth noting that the uncoated substrate was more affected by the corrosion than the coated surfaces. The destruction of the protection layer can be seen on the coated samples, after performing the corrosion tests. Regarding the TiCN coatings, there were various corrosion products on the surface, indicating that this surface was affected by the corrosive solution. Moreover, the coating was partially destroyed in some areas, with the CoCr substrate being visible. The TiSiCN surface also had corrosion products, but there were less compared with the TiCN surface, indicating better anticorrosive properties. All surfaces were affected by the corrosive solution, but TiSiCN was less damaged.

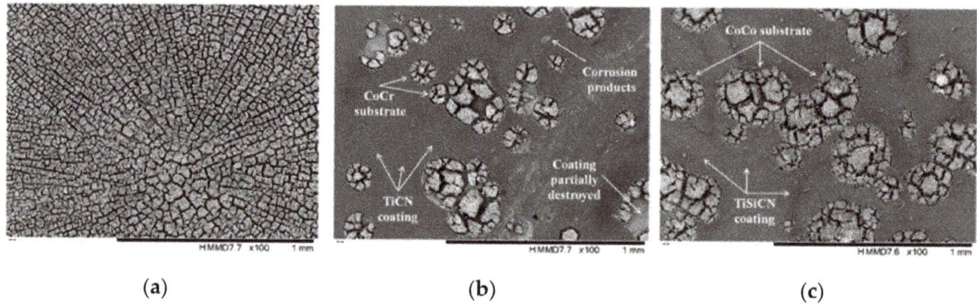

Figure 6. SEM of the investigated (**a**) CoCr substrate and (**b**) TiCN and (**c**) TiSiCN coatings.

The main roughness parameters of the investigated surfaces are presented in Table 6. Comparing these results with the values obtained before the corrosion tests (Table 3), it can be said that all surfaces were significantly damaged after the corrosion tests. The R_a of the uncoated substrates increased from 46.9 ± 5.9 nm to 1342.9 ± 192.4 nm. A significant increase in R_a roughness was found for the TiCN coatings after corrosion (15584.3 ± 7462.8 nm), indicating a major deterioration of the coatings after the corrosion tests. The TiSiCN coatings were also affected by the corrosive process, but they have finer irregularities than TiCN, demonstrating that the addition of Si led to an enhancement of anticorrosive properties. All surfaces showed a negative value of S_k after the corrosion tests, signifying that the surfaces were characterized by many valleys formed during the corrosive processes. The TiSiCN surface had fewer valleys that the TiCN surface, as shown by the smaller absolute value of S_k.

Table 6. Main roughness parameters of the investigated specimens after corrosion tests.

Sample	CoCr	TiCN	TiSiCN
R_a (nm)	1342.9 ± 192.4	15,584.3 ± 7462.8	6297.9 ± 1598.3
R_q (nm)	1758.2 ± 237.2	19,896.6 ± 6410.9	7480.5 ± 1567.1
S_k	−0.1 ± 0.2	−1.2 ± 0.8	−0.7 ± 0.4
Kurtosis	2.4 ± 0.3	1.8 ± 0.4	0.7 ± 0.3

4. Discussions

The nature of the electrolyte, scanning rate, temperature, impurities, anode material and surface state of the samples are a part of the parameters, which can influence the electrochemical reactions. For example, the surface texture of the working electrode (investigated sample) is one of the most important parameters, which has an influence on the Tafel slopes, and consequently on the corrosion rate. Surface roughness has an important effect on general or pitting corrosion and the nucleation of metastable pitting. Skewness and kurtosis were used to identify the corrosion mechanism. Reid et al.

have patented a method and apparatus for the identification of corrosion in metal objects and defining typical values of skewness and kurtosis for the identification of corrosion mechanism [59]. The pitting mechanism appears when the $S_k < -2$. In all our cases, the S_k had values greater than −2, indicating that a general corrosion mechanism can be found for all investigated surfaces. Regarding the kurtosis, if the value is < 3, general corrosion can be observed, while for kurtosis > 3, pitting corrosion can be found. Considering this, it can be seen that for all investigated surfaces, a general corrosion mechanism has been identified. If the surface is rough, then a larger area could contribute to the increase in the corrosion current or to the corrosion rate. Therefore, a decreased surface roughness will lead to a better corrosion resistance [60]. According to this statement, the TiCN-coated surface was rougher than TiSiCN, and this is probably the reason for its better corrosion resistance. For the uncoated CoCr, the roughness is not a factor which has a major influence on the corrosion resistance. Compared to the coated samples, the CoCr uncoated substrate had a smaller roughness, while its corrosion resistance was worse than for both coatings. Clearly, the surface roughness affects the corrosive behavior of materials (i.e., metals, alloys, coatings) and the nature of its effects (increase or decrease in the degree of corrosion) depends on the nature of the material.

To the best of our knowledge, there is no direct relation between hardness and corrosion resistance. Hard coatings can be subjected to surface microcracks, and then a localized penetration of corrosive solution will take place, leading to a galvanic cell, which accelerates the corrosive process. Hardness is important for load-bearing implants because a hardened material can have the ability to withstand wear. Taking into account the results of the present study, TiSiCN was harder than CoCr and the TiCN coatings and was more adequate for the proposed application.

For load-bearing implants, resistance to plastic deformation is an important factor and it can be described by the H^3/E^2 ratio. Moreover, a material with a high H^3/E^2 ratio resists plastic deformation during low load contact events and exhibits a higher yield strength [61,62]. It is also generally accepted that the H/E ratio can be considered an important indicator of a good wear resistance of the surface [63–65]. Thus, the improvement of the H/E ratio and, consequently, of the resistance to plastic deformation (H^3/E^2 ratio) of the load-bearing implant may offer advantages, such as less surface damage and increased durability. In this study, the TiSiCN coatings have an H^3/E^2 ratio equal to 1.1, which is higher than the one for TiCN (0.63), indicating that TiSiCN has a superior toughness and it can offer a better resistance to plastic deformation and good wear resistance.

The addition of Si to TiCN coatings leads to a grain refinement, and the crystallite size (d) was decreased to 16.4 nm in the case TiCN and to 14.6 nm in the case of TiSiCN. The formation of new defects, especially dislocations, is also responsible for the reduction in the crystallite size. The strain in the TiSiCN coatings ($\varepsilon = 0.012$) was lower compared to the TiCN coatings ($\varepsilon = 0.053$). The reason for this decrease could be attributed to different factors. One reason could be due to the addition of Si, which has atomic radii (0.111) smaller than that of Ti (0.146 nm), leading to a disorder of the crystal lattice, which is also evident by XRD diffraction (peaks were shifted when compared with the TiCN standard). The second reason could be attributed to amorphous phases in which Si can be found (Si, SiNx, SiCN) or C=C phases at grain boundaries, which are detrimental for crystallite development. TiSiCN has a higher C content than TiSiC. It is difficult to separate these factors and to know their contribution. However, the crystalline disorder becomes more pronounced by an increase in carbon content, which is also suggested by the decrease in the crystallite size and by the decrease in microstrain. Moreover, Franceschini et al. reported a strong dependence of stress on the nitrogen content in a-C:H films; at a low N content, the stress is high [66]. This effect is difficult to see in our coatings, because the N content is reduced after the addition of Si, but it is a minor reduction. This result can also have a major influence on the corrosion resistance of TiSiCN. This coating probably presents fewer defects and it is more compact than TiCN.

When the crystallite size decreases, the corrosion current density decreases and polarization resistance increases, which means that the corrosion resistance of the coatings increases with decreasing grain size. Thus, the TiSiCN coatings, which have the smallest crystalline size, were more resistant to corrosive attack. The dependence of corrosion resistance on crystallite size can be ascribed to the BOLS mechanism [67]. In the grain boundaries, there are undercoordinated atoms with lowered residual cohesive energy which possess high energy, these atoms exist in unstable states and an increase in their percent will lead to an increase of corrosion resistance [68]. This finding is also sustained by the strain ε value. In both cases, the strain was low, but after the addition of Si, the strain was significantly reduced. When the strain decreases, the corrosion resistance of the coatings increases. Thus, the correlation between high corrosion resistance and low strain and small crystallites can be explained in terms of the "bond-order-length-strength correlation mechanism", meaning that the undercoordinated atoms found on the surface or in grain boundaries take the responsibility of the good corrosion resistance. In the current paper, along with the addition of Si, the Q_{dl} parameter was also decreasing, and this result could be due to a smaller crystallite size obtained by the TiSiCN coating. Thus, the decreased generation of defects, in the case of this coating, had a beneficial effect on the protective properties. It was shown that defects within a structure can cause localized corrosion at the coating–substrate interface, due to the electrolyte ingress [69]. In addition, R_{pore} indicated that the resistivity of the electrolyte in the pores had the highest value in this case, which can be also correlated with the lack of defects. The α values ranged from about 0.87 to 0.90, and the deviation from an ideal capacitor was ascribed to differences in roughness, as was shown. The dependence between roughness, capacitance and associated α values was demonstrated [69], although there are also some other factors which can be influences, such as thickness and the dielectric constant of the material.

5. Conclusions

The present study aimed to find a new and improved possible solution for load-bearing implants. For this purpose, titanium-based carbonitrides with and without the addition of Si were investigated and compared. Both coatings obtained by the cathodic arc deposition method had an almost stoichiometric structure, being solid solution, which consisted of a mixture of TiC and TiN, with a face-centered cubic (FCC) structure. The crystallite size decreased with the addition of Si into the TiCN matrix, the crystallite size of TiCN was 16.4 nm, while for TiSiCN it was 14.6 nm. Both coated surfaces exhibited a uniform coverage, with some microparticles from the ion sputtering and ejection of the particles during the deposition process. After the addition of Si into TiCN, the R_a roughness values decreased, indicating a beneficial effect of Si.

All investigated surfaces had positive skewness, which is adequate for load-bearing implants, which work in corrosive environments. The hardness of the TiCN coating was 36.6 ± 2.9 GPa and it was significantly increased to 47.4 ± 1 GPa when small amounts of Si were added into the TiCN layer structure, while the elastic modulus was increased from 277 ± 8 GPa to 310 ± 2 GPa. A significant increase in resistance to plastic deformation (H^3/E^2 ratio) from 0.63 to 1.1 was found after the addition of Si into the TiCN matrix.

The most electropositive value of corrosion potential was found for the TiSiCN coating (−14 mV), as well as the smallest value of corrosion current density (49.6 nA cm^2), indicating good corrosion resistance. The TiSiCN coating exhibited the best protection after immersion in 90% DMEM + 10% FBS, the best capacitive character, indicated by the low value of Q_{dl}, and the highest resistance through the pores generated by the defects of the coatings and the electrolyte ingress, indicated by R_{pore} and R_{ct}.

According to the conducted research, TiSiCN coatings have shown good mechanical properties and high corrosion resistance and are a good alternative for the coating of load-bearing implants.

Author Contributions: Conceptualization, A.V. and C.V.; methodology, A.V.; investigation, M.D., P.S., I.G.S., and I.P.; resources, A.V.; data curation, I.P.; writing—original draft preparation, M.D., I.P., and P.S.; writing—review and editing, C.V. and A.V.; supervision, A.V. All authors have read and agreed to the published version of the manuscript.

Funding: The present work was supported under a grant of the Romanian National Authority for Scientific Research, CNCS—UEFISCDI, project No. PN-III-P1-1.2-PCCDI-2017-0239/60PCCDI 2018, within PNCDI III. The EDS, SEM, XRD and nanoindentation results were acquired using the systems purchased by the infrastructure project INOVA-OPTIMA SMIS code 49164, contract No. 658/2014. A part of work is also supported by PROINSTITUTIO Project–contract No. 19PFE/17.10.2018.

Conflicts of Interest: The authors declare no conflict of interest. The funders had no role in the design of the study; in the collection, analyses, or interpretation of data; in the writing of the manuscript, or in the decision to publish the results.

Nomenclature

Roman Symbols	
CPE_{dl}	Constant phase element that models the behavior of a double layer
CPE_{layer}	Constant phase element that models the behavior of a layer
d	Crystallite size
E	Elastic modulus
E_{corr}	Corrosion potential
E_{OC}	Open circuit potential
F	Force
H	Hardness
i_{corr}	Corrosion current density
Q_{dl}	Capacitance of the double layer
Q_{layer}	Capacitance of the layer
R_a	Arithmetic average of the roughness profile
R_{ct}	Charge transfer resistance
R_{el}	Electrolyte resistance
RMS	Root mean square
R_p	Polarization resistance
R_{pore}	Resistance associated to the current flow through the pores
R_q	Root means square average of the roughness profile
S_k	Skewness
T(hkl)	Texture coefficient
Z	Impedance
Z_{im}	Imaginary part of impedance
Z_{re}	Real part of impedance
Greek Symbols	
2θ	Angle between incident beam and reflected beam
α dl	Exponent equaling 1 for a capacitor characteristic to the double layer
α layer	Exponent equaling 1 for a capacitor characteristic to the layer
β_a	Anodic Beta coefficient of Tafel slope
β_c	Cathodic Beta coefficient of Tafel slope
ε	Strain
χ^2	Chi-square statistic distribution
Acronyms	
CE	Counter electrode
CVD	Chemical vapor deposition
DMEM	Dulbecco's Modified Eagle's Medium
EDS	Energy dispersive spectrometry
EIS	Electrochemical impedance spectroscopy
FBS	Fetal bovine serum
FCC	Face centered cubic structure
JCPDS	Joint committee on powder diffraction standards
PVD	Physical vapor deposition
RE	Reference electrode
SCE	Saturated Calomel electrode
SEM	Scanning electron microscopy
XRD	X-Ray Diffraction

References

1. Garcia, J.A.; Diaz, C.; Mandl, S.; Lutz, J.; Martinez, R.; Rodriguez, R.J. Tribological improvements of plasma immersion implanted CoCr alloys. *Surf. Coat. Technol.* **2010**, *204*, 2928–2932. [CrossRef]
2. Okazaki, Y.; Gotoh, E. Comparison of metal release from various metallic biomaterials in vitro. *Biomaterials* **2005**, *26*, 11–21. [CrossRef] [PubMed]
3. Sahoo, P.; Das, S.K.; Paulo Davim, J. 1—Tribology of materials for biomedical applications. In *Mechanical Behaviour of Biomaterials*; Davim, J.P., Ed.; Woodhead Publishing: Cambridge, UK, 2019; pp. 1–45. [CrossRef]
4. Goharian, A.; Kadir, M.R.A.; Abdullah, M.R. *Trauma Plating Systems: Biomechanical, Material, Biological, and Clinical Aspects*; Elsevier Science: Amsterdam, The Netherlands, 2017.
5. Lamerain, M.; Bachy, M.; Delpont, M.; Kabbaj, R.; Mary, P.; Vialle, R. CoCr rods provide better frontal correction of adolescent idiopathic scoliosis treated by all-pedicle screw fixation. *Eur. Spine J.* **2014**, *23*, 1190–1196. [CrossRef] [PubMed]
6. Serhan, H.; Mhatre, D.; Newton, P.; Giorgio, P.; Sturm, P. Would CoCr Rods Provide Better Correctional Forces Than Stainless Steel or Titanium for Rigid Scoliosis Curves? *J. Spinal Disord. Tech.* **2013**, *26*, E70–E74. [CrossRef] [PubMed]
7. Hanawa, T. 1—Overview of metals and applications. In *Metals for Biomedical Devices*; Niinomi, M., Ed.; Woodhead Publishing: Cambridge, UK, 2010; pp. 3–24. [CrossRef]
8. Minnath, M.A. 7—Metals and alloys for biomedical applications. In *Fundamental Biomaterials: Metals*; Balakrishnan, P., Sreekala, M.S., Thomas, S., Eds.; Woodhead Publishing: Cambridge, UK, 2018; pp. 167–174. [CrossRef]
9. Nine, M.J.; Choudhury, D.; Hee, A.C.; Mootanah, R.; Abu Osman, N.A. Wear Debris Characterization and Corresponding Biological Response: Artificial Hip and Knee Joints. *Materials* **2014**, *7*, 980–1016. [CrossRef] [PubMed]
10. Liu, H.T.; Ge, S.R.; Cao, S.F.; Wang, S.B. Comparison of wear debris generated from ultra high molecular weight polyethylene in vivo and in artificial joint simulator. *Wear* **2011**, *271*, 647–652. [CrossRef]
11. Mattei, L.; Di Puccio, F.; Piccigallo, B.; Ciulli, E. Lubrication and wear modelling of artificial hip joints: A review. *Tribol. Int.* **2011**, *44*, 532–549. [CrossRef]
12. Fujiwara, N.; Kobayashi, K. Macrophages in inflammation. *Curr Drug Targets Inflamm Allergy* **2005**, *4*, 281–286. [CrossRef]
13. Man, K.; Jiang, L.H.; Foster, R.; Yang, X.B.B. Immunological Responses to Total Hip Arthroplasty. *J. Funct. Biomater.* **2017**, *8*, 33. [CrossRef]
14. Perez-Maceda, B.T.; Lopez-Fernandez, M.E.; Diaz, I.; Kavanaugh, A.; Billi, F.; Escudero, M.L.; Garcia-Alonso, M.C.; Lozano, R.M. Macrophage Biocompatibility of CoCr Wear Particles Produced under Polarization in Hyaluronic Acid Aqueous Solution. *Materials* **2018**, *11*, 756. [CrossRef]
15. Hauer, G.; Leitner, L.; Ackerl, C.M.; Klim, S.; Vielgut, I.; Ehall, R.; Glehr, M.; Leithner, A.; Sadoghi, P. Titanium-Nitride Coating Does Not Result in a Better Clinical Outcome Compared to Conventional Cobalt-Chromium Total Knee Arthroplasty after a Long-Term Follow-Up: A Propensity Score Matching Analysis. *Coatings* **2020**, *10*, 442. [CrossRef]
16. Eltit, F.; Assiri, A.; Garbuz, D.; Duncan, C.; Masri, B.; Greidanus, N.; Bell, R.; Sharma, M.; Cox, M.; Wang, R.Z. Adverse reactions to metal on polyethylene implants: Highly destructive lesions related to elevated concentration of cobalt and chromium in synovial fluid. *J. Biomed. Mater. Res. Part A* **2017**, *105*, 1876–1886. [CrossRef]
17. José, A.; García, P.J.R.; Rocío, O.; Iban, Q.; Rafael, J. Rodríguez. Advanced Surface Treatments for Improving the Biocompatibility of Prosthesis and Medical Implants. In *Advanced Surface Engineering Research*; Chowdhury, M.A., Ed.; IntechOpen: London, UK, 2018. [CrossRef]
18. Anders, A. Energetic deposition using filtered cathodic arc plasmas. *Vacuum* **2002**, *67*, 673–686. [CrossRef]
19. Sanders, D.M.; Anders, A. Review of cathodic arc deposition technology at the start of the new millennium. *Surf. Coat. Technol.* **2000**, *133*, 78–90. [CrossRef]
20. Setsuhara, Y. 4.12—Plasma Sources in Thin Film Deposition. In *Comprehensive Materials Processing*; Hashmi, S., Batalha, G.F., Van Tyne, C.J., Yilbas, B., Eds.; Elsevier: Oxford, UK, 2014; pp. 307–324. [CrossRef]

21. Karlsson, L.; Hultman, L.; Johansson, M.P.; Sundgren, J.E.; Ljungcrantz, H. Growth, microstructure, and mechanical properties of arc evaporated TiCxN1-x (0 <= x <= 1). *Surf. Coat. Technol.* **2000**, *126*, 1–14. [CrossRef]
22. Liskiewicz, T.; Rybiak, R.; Fouvry, S.; Wendler, B. Fretting wear of Ti(CxNy) PVD coatings under variable environmental conditions. *Proc. Inst. Mech. Eng. Part J-J. Eng. Tribol.* **2006**, *220*, 125–134. [CrossRef]
23. Lotfi, M.; Ashrafi, H.; Amini, S.; Farid, A.A.; Jahanbakhsh, M. Characterization of various coatings on wear suppression in turning of Inconel 625: A three-dimensional numerical simulation. *Proc. Inst. Mech. Eng. Part J-J. Eng. Tribol.* **2017**, *231*, 734–744. [CrossRef]
24. Chandrashekharaiah, T.M.; Kori, S.A. Effect of grain refinement and modification on the dry sliding wear behaviour of eutectic Al-Si alloys. *Tribol. Int.* **2009**, *42*, 59–65. [CrossRef]
25. Shtansky, D.V.; Gloushankova, N.A.; Sheveiko, A.N.; Kiryukhantsev-Korneev, P.V.; Bashkova, I.A.; Mavrin, B.N.; Ignatov, S.G.; Filippovich, S.Y.; Rojas, C. Si-doped multifunctional bioactive nanostructured films. *Surf. Coat. Technol.* **2010**, *205*, 728–739. [CrossRef]
26. Shi, Z.F.; Wang, Y.J.; Du, C.; Huang, N.; Wang, L.; Ning, C.Y. Silicon nitride films for the protective functional coating: Blood compatibility and biomechanical property study. *J. Mech. Behav. Biomed. Mater.* **2012**, *16*, 9–20. [CrossRef]
27. Endler, I.; Hohn, M.; Schmidt, J.; Scholz, S.; Herrmann, M.; Knaut, M. Ternary and quarternary TiSiN and TiSiCN nanocomposite coatings obtained by Chemical Vapor Deposition. *Surf. Coat. Technol.* **2013**, *215*, 133–140. [CrossRef]
28. Hatem, A.; Lin, J.L.; Wei, R.H.; Torres, R.D.; Laurindo, C.; de Souza, G.B.; Soares, P. Tribocorrosion behavior of low friction TiSiCN nanocomposite coatings deposited on titanium alloy for biomedical applications. *Surf. Coat. Technol.* **2018**, *347*, 1–12. [CrossRef]
29. Johnson, L.J.S.; Rogstrom, L.; Johansson, M.P.; Oden, M.; Hultman, L. Microstructure evolution and age hardening in (Ti,Si)(C,N) thin films deposited by cathodic arc evaporation. *Thin Solid Film.* **2010**, *519*, 1397–1403. [CrossRef]
30. Wei, R.H. Plasma enhanced magnetron sputter deposition of Ti-Si-C-N based nanocomposite coatings. *Surf. Coat. Technol.* **2008**, *203*, 538–544. [CrossRef]
31. Guruvenket, S.; Li, D.; Klemberg-Sapieha, J.E.; Martinu, L.; Szpunar, J. Mechanical and tribological properties of duplex treated TiN, nc-TiN/a-SiNx and nc-TiCN/a-SiCN coatings deposited on 410 low alloy stainless steel. *Surf. Coat. Technol.* **2009**, *203*, 2905–2911. [CrossRef]
32. Ma, S.L.; Ma, D.Y.; Guo, Y.; Xu, B.; Wu, G.Z.; Xu, K.W.; Chu, P.K. Synthesis and characterization of super hard, self-lubricating Ti-Si-C-N nanocomposite coatings. *Acta Mater.* **2007**, *55*, 6350–6355. [CrossRef]
33. Constantin, L.R.; Parau, A.C.; Balaceanu, M.; Dinu, M.; Vladescu, A. Corrosion and tribological behaviour in a 3.5% NaCl solution of vacuum arc deposited ZrCN and Zr-Cr-Si-C-N coatings. *Proc. Inst. Mech. Eng. Part J-J. Eng. Tribol.* **2019**, *233*, 158–169. [CrossRef]
34. Wang, Y.; Li, J.L.; Dang, C.Q.; Wang, Y.X.; Zhu, Y.J. Influence of bias voltage on structure and tribocorrosion properties of TiSiCN coating in artificial seawater. *Mater. Charact.* **2017**, *127*, 198–208. [CrossRef]
35. Sundgren, J.E.; Johansson, B.O.; Hentzell, H.T.G.; Karlsson, S.E. Mechanisms of reactive sputtering of titanium nitride and titanium carbide.3. Influence of substrate bias on composition and structure. *Thin Solid Film.* **1983**, *105*, 385–393. [CrossRef]
36. Saoula, N.; Henda, K.; Kesri, R. Effect Of The Plasma Deposition Parameters On The Properties Of Ti/TiN Multilayers For Hard Coatings Applications. In Proceedings of the 1st International Conference on Laser and Plasma Applications in Materials Science, Algiers, Algeria, 23–26 June 2008; p. 256.
37. Constantin, L.; Braic, M.; Dinu, M.; Balaceanu, M.; Braic, V.; Farcau, C.; Vladescu, A. Effects of Zr, Nb, or Si addition on the microstructural, mechanical, and corrosion resistance of TiCN hard coatings. *Mater. Corros. -Werkst. Und Korros.* **2016**, *67*, 929–938. [CrossRef]
38. Ritchie, R.O. The conflicts between strength and toughness. *Nat. Mater.* **2011**, *10*, 817–822. [CrossRef] [PubMed]
39. Grieveson, P. An investigation of the Ti-C-N system. *Proc. Br. Ceram. Soc.* **1967**, *8*, 137–153.
40. Terry, B.S.; Grieveson, P. Observations on the Role of Interfacial Phenomena in Materials Processing. *ISIJ Int.* **1993**, *33*, 166–175. [CrossRef]
41. Levi, G.; Kaplan, W.D.; Bamberger, M. Structure refinement of titanium carbonitride (TiCN). *Mater. Lett.* **1998**, *35*, 344–350. [CrossRef]

42. Nietubyc, R.; Lorkiewicz, J.; Sekutowicz, J.; Smedley, J.; Kosinska, A. Optimization of cathodic arc deposition and pulsed plasma melting techniques for growing smooth superconducting Pb photoemissive films for SRF injectors. *Nucl. Instrum. Methods Phys. Res. Sect. A-Accel. Spectrometers Detect. Assoc. Equip.* **2018**, *891*, 78–86. [CrossRef]
43. Panjan, P.; Drnovšek, A.; Gselman, P.; Čekada, M.; Panjan, M. Review of Growth Defects in Thin Films Prepared by PVD Techniques. *Coatings* **2020**, *10*, 447. [CrossRef]
44. ISO 2517:1974 Diacetone Alcohol for Industrial Use–List of Methods of Test. Edition No. 3, Publication Date: 1974–04. Available online: https://www.iso.org/standard/7452.html (accessed on 21 May 2020).
45. Schuh, C.A. Nanoindentation studies of materials. *Mater. Today* **2006**, *9*, 32–40. [CrossRef]
46. Fujisawa, N.; Lukomski, M. Nanoindentation near the edge of a viscoelastic solid with a rough surface. *Mater. Des.* **2019**, *184*. [CrossRef]
47. Fischer-Cripps, A.C. Critical review of analysis and interpretation of nanoindentation test data. *Surf. Coat. Technol.* **2006**, *200*, 4153–4165. [CrossRef]
48. Pana, I.; Vitelaru, C.; Kiss, A.; Zoita, N.C.; Dinu, M.; Braic, M. Design, fabrication and characterization of TiO2-SiO2 multilayer with tailored color glazing for thermal solar collectors. *Mater. Des.* **2017**, *130*, 275–284. [CrossRef]
49. Martinez, E.; Romero, J.; Lousa, A.; Esteve, J. Nanoindentation stress-strain curves as a method for thin-film complete mechanical characterization: Application to nanometric CrN/Cr multilayer coatings. *Appl. Phys. A-Mater. Sci. Process.* **2003**, *77*, 419–426. [CrossRef]
50. Shan, L.; Wang, Y.X.; Li, J.L.; Li, H.; Wu, X.D.; Chen, J.M. Tribological behaviours of PVD TiN and TiCN coatings in artificial seawater. *Surf. Coat. Technol.* **2013**, *226*, 40–50. [CrossRef]
51. Li, W.; Liu, P.; Xue, Z.H.; Ma, F.C.; Zhang, K.; Chen, X.H.; Feng, R.; Liaw, P.K. Microstructures, mechanical behavior and strengthening mechanism of TiSiCN nanocomposite films. *Sci. Rep.* **2017**, *7*. [CrossRef]
52. Musil, J.; Jilek, R.; Meissner, M.; Tolg, T.; Cerstvy, R. Two-phase single layer Al-O-N nanocomposite films with enhanced resistance to cracking. *Surf. Coat. Technol.* **2012**, *206*, 4230–4234. [CrossRef]
53. Musil, J.; Jirout, M. Toughness of hard nanostructured ceramic thin films. *Surf. Coat. Technol.* **2007**, *201*, 5148–5152. [CrossRef]
54. Pauleau, Y.; Thiery, E. Deposition and characterization of nanostructured metal/carbon composite films. *Surf. Coat. Technol.* **2004**, *180*, 313–322. [CrossRef]
55. Cordoba-Torres, P. Relationship between constant-phase element (CPE) parameters and physical properties of films with a distributed resistivity. *Electrochim. Acta* **2017**, *225*, 592–604. [CrossRef]
56. Hirschorn, B.; Orazem, M.E.; Tribollet, B.; Vivier, V.; Frateur, I.; Musiani, M. Determination of effective capacitance and film thickness from constant-phase-element parameters. *Electrochim. Acta* **2010**, *55*, 6218–6227. [CrossRef]
57. Carlos Valero, V.; Anna Igual, M. *Electrochemical Aspects in Biomedical Alloy Characterization: Electrochemical Impedance Spectrosopy*; IntechOpen: London, UK, 2011. [CrossRef]
58. Dinu, M.; Hauffman, T.; Cordioli, C.; Vladescu, A.; Braic, M.; Hubin, A.; Cotrut, C.M. Protective performance of Zr and Cr based silico-oxynitrides used for dental applications by means of potentiodynamic polarization and odd random phase multisine electrochemical impedance spectroscopy. *Corros. Sci.* **2017**, *115*, 118–128. [CrossRef]
59. Reid, S.A.; Eden, D.A. Assessment of Corrosion. U.S. Patent No. 6,264,824, 24 July 2001.
60. Toloei, A.a.S.V.a.N.D. The Relationship Between Surface Roughness and Corrosion. *Asme Int. Mech. Eng. Congr. Expo. Proc. (IMECE)* **2013**, *2*. [CrossRef]
61. Musil, J. Hard and superhard nanocomposite coatings. *Surf. Coat. Technol.* **2000**, *125*, 322–330. [CrossRef]
62. Roy, M.E.; Whiteside, L.A.; Xu, J.; Katerberg, B.J. Diamond-like carbon coatings enhance the hardness and resilience of bearing surfaces for use in joint arthroplasty. *Acta Biomater.* **2010**, *6*, 1619–1624. [CrossRef] [PubMed]
63. Leyland, A.; Matthews, A. On the significance of the H/E ratio in wear control: A nanocomposite coating approach to optimised tribological behaviour. *Wear* **2000**, *246*, 1–11. [CrossRef]
64. Ievlev, V.M.; Kostyuchenko, A.V.; Darinskii, B.M.; Barinov, S.M. Hardness and microplasticity of nanocrystalline and amorphous calcium phosphate coatings. *Phys. Solid State* **2014**, *56*, 321–329. [CrossRef]

65. Surmeneva, M.A.; Tyurin, A.I.; Teresov, A.D.; Koval, N.N.; Pirozhkova, T.S.; Shuvarin, I.A.; Surmenev, R.A. Combined effect of pulse electron beam treatment and thin hydroxyapatite film on mechanical features of biodegradable AZ31 magnesium alloy. In Proceedings of the 3rd International Youth Conference on Interdisciplinary Problems of Nanotechnology, Biomedicine and Nanotoxicology (Nanobiotech), Tambov, Russia, 21–22 May 2015.
66. Franceschini, D.F.; Achete, C.A.; Freire, F.L. Internal-Stress Reduction by Nitrogen Incorporation in Hard Amorphous-Carbon Thin-Films. *Appl. Phys. Lett.* **1992**, *60*, 3229–3231. [CrossRef]
67. Sun, C.Q. Size dependence of nanostructures: Impact of bond order deficiency. *Prog. Solid State Chem.* **2007**, *35*, 1–159. [CrossRef]
68. Sun, C.Q.; Shi, Y.; Li, C.M.; Li, S.; Yeung, T.C.A. Size-induced undercooling and overheating in phase transitions in bare and embedded clusters. *Phys. Rev. B* **2006**, *73*. [CrossRef]
69. Liu, C.; Bi, Q.; Leyland, A.; Matthews, A. An electrochemical impedance spectroscopy study of the corrosion behaviour of PVD coated steels in 0.5 N NaCl aqueous solution: Part II. EIS interpretation of corrosion behaviour. *Corros. Sci.* **2003**, *45*, 1257–1273. [CrossRef]

Article

Production of High Silicon-Doped Hydroxyapatite Thin Film Coatings via Magnetron Sputtering: Deposition, Characterisation, and In Vitro Biocompatibility

Samuel C. Coe [1], Matthew D. Wadge [1,*], Reda M. Felfel [1,2], Ifty Ahmed [1], Gavin S. Walker [1], Colin A. Scotchford [1] and David M. Grant [1,*]

1 Advanced Materials Research Group, Faculty of Engineering, University of Nottingham, Nottingham NG7 2RD, UK; samuelccoe@gmail.com (S.C.C.); Reda.Felfel@nottingham.ac.uk (R.M.F.); ifty.ahmed@nottingham.ac.uk (I.A.); Gavin.Walker@nottingham.ac.uk (G.S.W.); colin.scotchford@nottingham.ac.uk (C.A.S.)
2 Physics Department, Faculty of Science, Mansoura University, Mansoura 35511, Egypt
* Correspondence: matthew.wadge@nottingham.ac.uk (M.D.W.); david.grant@nottingham.ac.uk (D.M.G.)

Received: 20 January 2020; Accepted: 20 February 2020; Published: 23 February 2020

Abstract: In recent years, it has been found that small weight percent additions of silicon to HA can be used to enhance the initial response between bone tissue and HA. A large amount of research has been concerned with bulk materials, however, only recently has the attention moved to the use of these doped materials as coatings. This paper focusses on the development of a co-RF and pulsed DC magnetron sputtering methodology to produce a high percentage Si containing HA (SiHA) thin films (from 1.8 to 13.4 wt.%; one of the highest recorded in the literature to date). As deposited thin films were found to be amorphous, but crystallised at different annealing temperatures employed, dependent on silicon content, which also lowered surface energy profiles destabilising the films. X-ray photoelectron spectroscopy (XPS) was used to explore the structure of silicon within the films which were found to be in a polymeric (SiO_2; Q^4) state. However, after annealing, the films transformed to a SiO_4^{4-}, Q^0, state, indicating that silicon had substituted into the HA lattice at higher concentrations than previously reported. A loss of hydroxyl groups and the maintenance of a single-phase HA crystal structure further provided evidence for silicon substitution. Furthermore, a human osteoblast cell (HOB) model was used to explore the in vitro cellular response. The cells appeared to prefer the HA surfaces compared to SiHA surfaces, which was thought to be due to the higher solubility of SiHA surfaces inhibiting protein mediated cell attachment. The extent of this effect was found to be dependent on film crystallinity and silicon content.

Keywords: hydroxyapatite; thin film; RF magnetron sputtering; pulsed DC; Silicon

---

1. Introduction

One of the many issues faced with regards to foreign objects being inserted into the body is the reduction or elimination of adverse effects caused by a stimulated immune response. Many such effects can be reduced with a careful material selection; however the current trend is to use the response of the body to improve the initial success of implant materials which could lead to a longer device lifetime, reducing the need for revision surgery. In the case of load bearing implants, such as in total hip replacement (THR) or total knee replacement (TKR), further complications such as the mechanical requirements of the material to bear load can reduce the number of materials available. It is often the case that materials that possess the mechanical properties are not always the most compatible in these sites. A possible solution is to coat mechanically sound devices with materials which will be more biologically favourable [1]. A considerable amount of work has already gone into this area [2,3] but many problems such as selection of the most appropriate correct coating technique is required to control properties such as coating adhesion [4], composition [5] and topography [6], which may ultimately influence the biological response of the surface.

While many material systems offer attractive properties for biomedical applications, HA has attracted much attention due to its excellent bioactive ability for the application of bone repair and bone replacement [7]. Unfortunately, the nature of ceramic materials due to their reduced mechanical performance in comparison to metals, means that they, including HA, are restricted to the application of non-load bearing sites [8]. This problem can be overcome by applying an HA coating onto metallic materials with superior mechanical strength and stiffness [9]. Currently, plasma spraying is the only suitable commercial and FDA approved coating process employed [10,11], but coatings are compromised by the inclusion of undesired calcium phosphate phases leading to bioresorption and variable tissue response [12]. In addition, these coatings are often poorly bonded to the metallic substrates, leading to failure in the form of cracking at the coating/substrate interface [13]. Physical vapour deposition (PVD), as an attractive alternative, offers dense, defect free, well adhered single phase coatings with controllable deposition parameters [14,15].

Numerous authors have showed RF magnetron sputtering could be a potentially beneficial technique for applying coatings to future implantable devices [16–20]. Current work has shown that HA coatings are bioactive and exhibit similar properties to bulk materials. Many different compositions have been achieved for HA thin films due to differing sputtering parameters [16–18,21]. However, it is still unclear what suitable deposition parameters are and how factors such as power density and sputtering environment affect resultant films. One important compositional parameter is the Ca/P ratio, which can give rise to changes in the thermal stability, biological response, mechanical properties and dissolution potential [17].

SiHA thin films have been shown to elicit an enhanced cellular response compared to HA, which may provide a way to increase bone response in vitro and in vivo, thus improving the longevity of implants for future generations [22,23]. Work concerning RF magnetron sputtered SiHA thin films have only recently been investigated in the last decade [19,23–26]. Thin film systems often differ from bulk material systems due to the manufacturing process. Therefore, it is assumed that silicate groups substitute for phosphate groups and it is not fully understood what may happen at higher silicon concentrations above 5 wt.% [23]. Furthermore, optimal silicon values (ca. 2.2 wt.%) have been suggested but these are not in agreement [27,28]. Therefore, higher silicon additions [29] up to 13.4 wt.% and altered crystallinities have been investigated in this study.

2. Materials and Methods

2.1. Substrate Preparation

A commercially pure grade 1 titanium sheet (Timet, Swansea, UK) (CPTi) was wire eroded into 10 mm diameter, 1 mm thick discs, which were then ground to a mirror finish ($R_a \sim 16 \pm 4$ nm) using a series of silicon carbide paper (P240-P1200; Struers, Rotherham, UK), polished with a mixture of colloidal silica (Buehler, Germany) and 10% hydrogen peroxide (Fisher Scientific, Loughborough, UK) and finally water. The discs were cleaned for 30 min in acetone, IMS and distilled water and dried under flowing nitrogen (BOC, London, UK).

Silicon single crystals (100) orientated Czochralski wafer (Compart Technology Ltd., Peterborough, UK) were cut into squares approximately 10 mm × 10 mm in size with a diamond tipped scribe, and were used as substrates for XRD to reduce peak contributions from the substrate.

2.2. Target Preparation

The HA targets were manufactured by plasma spraying HA powder (Plasma Biotal, Buxton, UK) onto a circular copper backing plate 75 mm in diameter and 5 mm in thickness, with a target material thickness of ca. 500 µm. For the Si doping, a single 99.999% pure silicon target (Kurt J. Lesker, Hastings, UK) was used, with dimensions 300 mm × 100 mm × 0.635 mm. Figure A2 demonstrates the target composition via XRD analysis to be mostly HA.

2.3. RF/Pulsed DC Magnetron Sputtering

The HA and SiHA thin film coatings were produced using a TEER UDP-650 type 2 unbalanced magnetron sputtering rig (Teer Coatings Ltd., Droitwich, UK). Before deposition, the chamber was pumped using successive rotary (Edwards E2M40 rotary pump; 10^{-2} Torr) and diffusion (Edwards 750 diffusion pump) pumping steps to a minimum of 2×10^{-5} Torr. The chamber was then backfilled with argon (36 sccm; 2×10^{-3} Torr; 99.999% purity; Pureshield BOC©, London, UK).

The HA target was powered by an Advanced Energies RF power supply unit whilst the silicon target was powered by an Advanced Energy Inc. (Bicester, UK) DC Pinnacle pulsed DC power supply. The frequency and pulse time were maintained constant at 150 kHz and 1500 ns, respectively. A −30 V bias was applied to the substrates. The samples were mounted on a stainless steel plate with double sided adhesive tape and rotated at 4 RPM. The samples were bias-cleaned for 2 min prior to sputtering. The total deposition time for each coating type was maintained at 2 h, with approximate sputtering rates of ca. 100 nm/h.

2.4. Post Deposition Heat Treatments

The obtained thin film coatings were recrystallised via heat treatment at a range of temperatures, with a set time of 2 h using a Lenton Thermal Design tube furnace (Hope Valley, UK) in an argon atmosphere. Argon was flowed at 100 sccm for 30 min before ramping the temperature up to 20 °C min^{-1}. Mass spectrometry indicated that all gases and water vapour were at untraceable levels before ramping. The samples were then left to cool under flowing argon.

2.5. Materials Characterisation

2.5.1. Scanning Electron Microscopy (SEM) and Energy Dispersive X-Ray Spectroscopy (EDX)

A Phillips XL-30 scanning electron microscope (LaB_6, Surrey, UK) was used to obtain micrographs at accelerating voltages between 10–20 keV and a working distance of 10 mm. An Oxford Instruments energy dispersive X-ray microanalysis (EDX) system (High Wycombe, UK) was utilised with a collecting time of 50 s. Ten randomly selected spots were analysed. Approximately 90,000 total spectrum counts were taken per sample area.

2.5.2. Focused Ion Beam SEM (FIB-SEM) and Transmission Electron Microscopy (TEM)

The HA and SiHA thin films were initially sectioned using a FEI Quanta 200 3D FIB-SEM (Cambridge, UK) fitted with a Quorum cryo-transfer unit, an Omniprobe micromanipulator and an INCA Oxford Instruments EDX analysis system. The FIB-SEM was operated at an ion beam accelerating voltage of 30 kV and electron beam accelerating voltages of 5–20 kV. A coating of tungsten was deposited in situ by chemical vapour deposition (CVD) to protect the HA coating. FEI-Runscript software was employed to mill inspection trenches, using reducing milling currents of 7 to 1 nA for rough sectioning, followed by milling currents of 0.5 nA to 30 pA in order to polish the lamella surfaces and to fashion U-shaped cuts into the lamella to facilitate lift-out. A cross sectional microstructural and chemical analysis was performed using a JEOL 2000-FX-II TEM (Welwyn Garden City, UK) operating at 200 keV.

2.5.3. X-Ray Diffraction (XRD)

A Bruker AXS D8 Advance X-ray diffractometer (Coventry, UK) was used in glancing angle X-ray diffraction mode. Cu kα X-rays (λ = 1.5406 Å) incidented the samples at a 2θ range of 20°–55°, a step size of 0.02° and a dwell time of 11 s. The samples were mounted and rotated at 30 RPM. A 2θ range and dwell time of 25°–35° and 2 s, respectively, was used for recrystallisation measurements. Temperature scans were performed from room temperature to a maximum of 800 °C to investigate the recrystallisation of thin films.

PC-APD version 3.5B DOS-based software was used to determine crystallite size of thin films using the Scherrer equation:

$$B = \frac{0.9\lambda}{t\cos\theta} \tag{1}$$

where B is the peak broadening (full width half maximum (FWHM) in radians), λ is the wavelength of the XRD source material and t is the crystallite diameter. B factors in instrumental broadening ($B = B_{Obs} - B_{inst}$) where B_{obs} is the observed line broadening, which includes instrumental factors such as detector slit width, area of the specimen irradiated and possible $K_{\alpha 2}$ X-rays.

2.5.4. Reflective High-Energy Electron Diffraction

A RHEED unit coupled with a JEOL 2000fx TEM operated at 200 keV was used to assess the surface crystallinity of films. The samples were held perpendicular to normal on a stage positioned immediately below the objective lens. The samples were tilted so that the shadow edge was positioned close to the primary beam to access near surface diffraction information. A GaN/GaAs single crystal standard reference sample with known *d*-spacings and a camera constant of 33.8 ± 0.5 cm was used to calculate thin films' *d*-spacings.

2.5.5. X-Ray Photoelectron Spectroscopy (XPS)

A Kratos Instruments Axis Ultra (Manchester, UK) with a monochromated Al Kα X-ray source was run at 10 keV and 15 mA. The instrument was operated in constant analysis energy (CAE) mode with a pass energy of 20 eV for high-resolution scans. Chemical and compositional information was obtained between 0–1400 eV. All samples were charge corrected to the C 1s adventitious carbon peak, which was set to a value of 284.8 eV. The region areas were selected manually and peak deconvolution was carried out using Gaussian-Lorentzian (GL30) line shapes. The data analysis and compositional quantification were carried out using CasaXPS software (version 2.3.22).

2.5.6. Fourier Transform Infrared Spectroscopy (FTIR)

Fourier transform infrared spectroscopy (FTIR) was performed on the thin films using a Bruker Optics Tensor 27 spectrometer (Coventry, UK) in glancing angle mode at 80° with a liquid nitrogen cooled MCT detector. The chamber was flushed with compressed air continually at 500 sccm to reduce background signal from water vapour. Samples scans were taken in triplicate for each coating with a total of 90 scans per specimen. Resultant spectra were analysed using an OPUS software (version 7.0).

2.5.7. Surface Roughness – Optical Profilometry

A Mitutoyo Surftest SV-600 profilometer (Coventry, UK) equipped with Surfpak SV software was used to measure surface roughness of the coatings. The stylus had a 5 μm radius tip. A scan length of 2 mm with a scan speed of 0.2 mm s^{-1} and a range of 80 μm were used for all samples. Calibration was carried out prior to every session using a Mitotoyo Precision reference specimen with an R_a value of 2.95 μm.

2.5.8. Surface Wettability—Sessile Drop/Contact Angle

Wetting angles of water droplets on the sample surfaces were tested using a sessile drop experiment. H_2O was pumped from a syringe at a rate of 1.0 $\mu L \cdot s^{-1}$, from a height 4.0 ± 0.5 mm above the sample's surface. A drop settling time of 10 s was implemented prior to data collection.

2.6. *In Vitro Biocompatibility*

2.6.1. Alamarblue™ Assay

Duplicate samples were placed in 24 well plates (Nunc, Warrington, UK) and UV sterilised. Tissue Culture Plastic (TCP) was used as a control surface. Human Osteoblasts (HOBs) from bone chips of femoral heads of patients undergoing total hip arthroplasty were seeded into wells at a density of 40,000 cells cm^{-2} and incubated at 37 °C and 5% CO_2. Media (Dulbecco/Vogt Modified Eagle's Minimal Essential Medium (DMEM) supplemented with 10% fetal bovine serum (Gibco Life Technologies, Inchinnan, Scotland), 1% L-Glutamine, 2% HEPES Buffer, 1% non-essential amino acids, 2% penicillin and streptomycin (Invitrogen, Rugby, UK) and 75 mg ascorbic acid (Sigma, UK)) were replenished every two days. At each respective time point (1, 4, 7, 10 and 14 days) the media were removed and samples were washed three times in phosphate-buffered saline (PBS) solution. Then, a 1 mL dilution of AlamarBlue™ (Serotec, Kidlington, UK) and Hank's balanced salt solution (HBSS) (Gibco, Inchinnan, Scotland) in the ratio of 1:10 was added to each well, including unseeded TCP wells, and incubated for 80 min. The well plates were subsequently wrapped in foil and shaken at 300 rpm on a Heidolph Titramax 100 for 10 min in a dark environment. The AlamarBlue™ solution was then removed and 100 μm aliquots were transferred into a 96 well plate (Nunc, Warrington, UK). Fluorescence was measured using a Bio Tek Instruments FLx800 fluorescence plate reader (Swindon, UK) at 560 nm excitation and 590 nm emission filters. Unreduced AlamarBlue™ was subtracted from recorded values to remove background signal. The experiments were repeated twice.

2.6.2. Alkaline Phosphatase Assay

After each designated time point, media were removed from culture plates and washed three times in PBS solution. Aliquots of 1 mL of sterile distilled water were added to each well. A freeze/thaw method was employed to lyse the cells. The samples were frozen at −20 °C and then allowed to defrost at room temperature. This was repeated three times. 50 μL of lysate solution was added to a 96 well plate per sample which was mixed with 50 μL of 4−nitrophenylphosphate (Sigma, UK) mixed with an appropriate quantity of diethanolamine buffer solution. Plates were shaken at 300 RPM for 1 min in a dark environment. The luminescence was then measured using a Bio Tek ELx800 luminescence plate reader with a primary wavelength of 405 nm and a reference wavelength of 630 nm.

2.6.3. DNA Hoechst Staining Assay

At each time point media were removed and HOBs were washed within PBS three times and submerged in 1 mL of sterile distilled water. A freeze/thaw cycle was carried out in triplicate to lyse HOB cell walls. Then, 100 μL of lysate was mixed with 100 μL of Hoechst 33,258 stain (Sigma, UK) and shaken at 300 RPM for 1 min in a dark environment. Fluorescence was then read on a Bio Tek Instruments FLx800 fluorescence plate reader with 360 nm excitation and 460 nm emission filters.

2.6.4. SEM Sample Preparation

At selected time points, media were removed and samples were washed with PBS thrice and replaced with 3% Glutaldehyde in 0.1 M sodium cacodylate buffer. This was replaced after 30 min with 7% sucrose solution in 0.1 M sodium cacodylate buffer. Specimens were then washed three times for 5 min periods with 0.1 M cacodylate buffer solution and then immersed in osmium tetraoxide for 1 h. Post fixing, cells were dehydrated using an ethanol/distilled water gradient (20% × 2 min, 40% × 5 min, 60% × 5 min, 80% × 5 min 90% × 5 min and 100% × 5 min × 2). Specimens were then submerged in hexamethyldisilazane (HMDS; Sigma, UK) for 5 min. This was then replaced with fresh HMDS and left to dry overnight. The samples were mounted on aluminium stubs with carbon adhesive tabs and gold/palladium coated (ca. 5 nm).

3. Results

3.1. Chemical and Structural Characterisation

3.1.1. Film Morphology and Thickness

Figure 1A shows a cross-section of a HA thin film on a CPTi disc annealed at 600 °C. Selected Area Electron Diffraction (SAED) confirmed that the substrate and the coating was indeed CPTi and polycrystalline HA. Moreover, the coating can be seen to be uniform in thickness, free of voids or defects and 185 ± 4 nm in thickness. Figure 1B shows that the SiHA3 thin film measured 216 ± 5 nm in thickness; films became thicker with increasing silicon content. Local recrystallisation of the HA films was noted, however, this may be a result of the e-beam interaction.

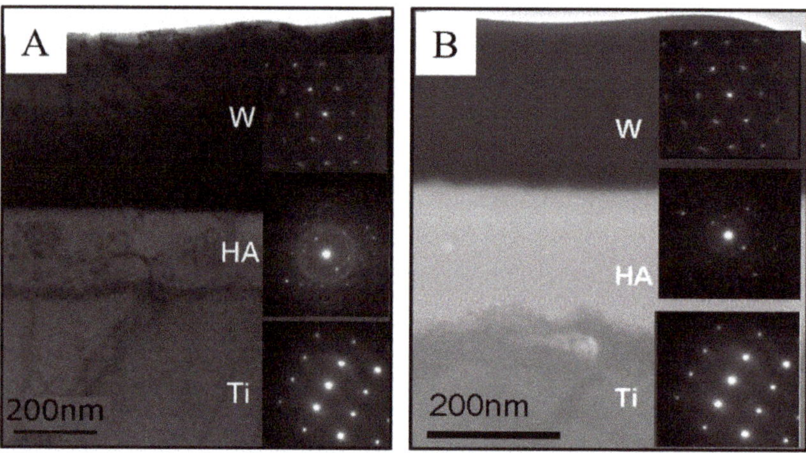

Figure 1. Bright field TEM image of (**A**) the tungsten/HA/Ti lamellar showing the HA thin film deposited on a CPTi substrate with associated selected area electron diffraction (SAED) patterns of the crystalline W protective coating, polycrystalline HA coating and crystalline Ti substrate and (**B**) the tungsten/SiHA3/Ti lamellar showing an amorphous SiHA3 coating on a CPTi substrate and associated SAED.

Surface micrographs of all films exhibited similar morphologies in as deposited and heat-treated states following the topography of the CPTi substrates. Figure 2A shows a representative as deposited HA film, showing a smooth dense coating without voids or defects. After a heat treatment at 600 °C (Figure 2B), all HA and silicon containing films looked similar in morphology with no notable differences with respect to silicon addition. At 700 °C, films became notably more textured with distinct features rising from the surfaces (Figure 2C).

Figure 2. Representative SEM micrographs displaying the surface morphology of HA and SiHA thin films where (**A**) is an as deposited thin film on a CPTi substrate, (**B**) annealed at 600 °C in flowing argon for 2 h and (**C**) 700 °C in flowing argon for 2 h. Images taken from HA samples.

3.1.2. Energy Dispersive X-Ray Analysis (EDX)

Table 1 demonstrates averaged Ca/P ratios of all coatings, and plasma sprayed targets. All samples were significantly higher than the stoichiometric value of bulk HA (1.67), however, they were significantly lower than the Ca/P ratio of the target material. The standard error of the mean was found to increase with increasing silicon content. Silicon content was seen to increase from 1.8 to 13.4 wt.%, with increasing target power densities from 6.6×10^{-4} to 3.3×10^{-3}, respectively.

Table 1. A summary of combined silicon content of HA and SiHA thin films batches as measured by EDX. Values displayed are the mean ± standard error of the mean, where $n = 6$. Ca/P content calculated from EDX and XPS measurements.

Sample	Silicon Target Density/Wcm^{-2}	Ca/P Ratio (EDX)	Ca/P Ratio (XPS)	Silicon Content/wt.%
Stoichiometric HA	N/A	1.67	1.67	N/A
Plasma Sprayed HA Target	N/A	1.90 ± 0.02	N/A	0.0 ± 0.1
HA	N/A	1.76 ± 0.03	1.43 ± 0.03	0.0 ± 0.1
SiHA1	6.6×10^{-4}	1.74 ± 0.03	1.23 ± 0.05	1.8 ± 0.3
SiHA2	1.6×10^{-3}	1.79 ± 0.08	1.16 ± 0.06	4.2 ± 0.7
SiHA3	3.3×10^{-3}	1.68 ± 0.09	1.03 ± 0.13	13.4 ± 1.4

3.1.3. X-Ray Diffraction (XRD) Analysis

XRD spectra (Figure 3) show representative XRD plots of as deposited films onto silicon (100) wafers heat treated at both 600 (Figure 3A) and 700 °C (Figure 3B). All as deposited (unannealed) coatings revealed an amorphous nature with a distinct hump at 27.5°. Following heat treatment at 600 °C, HA, SiHA1 and SiHA2 films recrystallised forming a single-phase HA structure matching ICDD card 09-432. Preferential orientation was seen along the (002) reflection when compared to a randomly orientated sample. The peak intensity of films decreased after inclusion of silicon. This was seen by peak broadening along the (002), (211), (112) and the (300) planes. Surprisingly, the SiHA3 samples remained amorphous after heat treatment at both 600 and 700 °C. The silicon addition to SiHA3 clearly had an effect on the recrystallisation transitional temperature of the HA structure. Full recrystallisation was not achieved for the SiHA2 at 700 °C with clear broadening of the FWHM compared to the full crystalline HA.

Consequently, a sequential heat treatment investigation (Figure 3C) was conducted on the SiHA3 coating to determine the temperature of recrystallisation. Figure 3C shows that the structure of SiHA3 did not alter after heat treatment up to 700 °C. At 800 °C, a single phase HA structure matching ICDD card number 09-432 was observed. Preferential orientation along the (002) plane was no longer observed, with the (211) plane being the most intense. For further heat treatments up to 1000 °C, the intensity of the (002) increased, and peaks sharpened, indicating crystal growth.

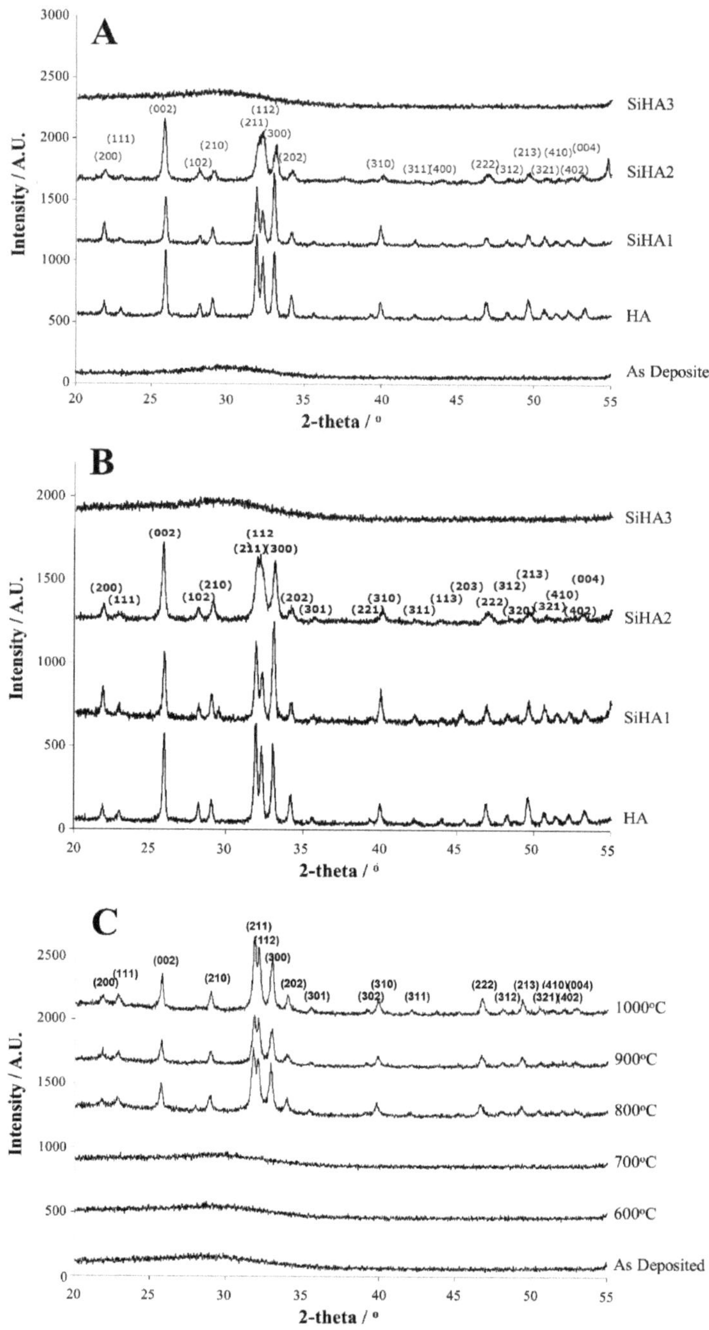

Figure 3. (**A**) and (**B**) Representative XRD plots of as deposited HA (unannealed), and HA and SiHA thin films on silicon (100) wafers heat treated at 600 °C and 700 °C in argon, respectively. (**C**) XRD plots of successive heat treatments on SiHA3 on silicon (100) wafer. Planes refer to ICDD 09-432 HA.

In addition, approximate crystallite size was calculated using the Scherrer equation. Table 2 lists calculated crystallite sizes of annealed HA and SiHA thin films from the XRD data presented. The crystallite size of all coatings increased with increasing annealing temperature. For example, the crystallite size of HA was ca. 78 and 89 nm following annealing at 600 and 700 °C, respectively. Respective values were seen to decrease with increasing silicon content. Crystallite size could not be calculated for SiHA3 at either annealing temperature, as both films were amorphous.

Table 2. A summary of HA crystallite size (nm) calculated by the Scherrer equation for HA and SiHA thin films sputtered onto silicon (100) single crystal wafer using the <001> planes.

Sample	**Heat Treatment Temperature/°C**	
	600	**700**
HA	78 ± 16	89 ± 16
SiHA1	65 ± 14	71 ± 14
SiHA2	71 ± 23	86 ± 31
SiHA3	N/A	N/A

3.1.4. RHEED Analysis

All as deposited samples were amorphous (not shown), with RHEED being conducted mainly to ascertain crystallinity and phase purity. After annealing at 600 °C, diffraction rings could be observed (Figure 4A–D). Due to the low accuracy of the RHEED measurements and the large number of HA diffractions, it proved difficult to index the rings. Therefore, only the most intense rings have been indexed with confidence, with the first ring corresponding to the (002) plane and the second broader ring corresponding to the (211), (112) and (300) planes. With increasing silicon content, the number of rings present decreased. The SiHA1 sample, Figure 4B, exhibited only two hazy rings corresponding to the d-spacings at 2.9 and 2.7 Å. The first ring was assigned to the (002) plane and the second broader ring is a combination of the (211), (112) and the (300) planes. The SiHA2 samples, Figure 4C, displayed the same rings as above but with lower intensity. The SiHA3 samples, however, showed no rings, indicating that these samples were amorphous.

The samples annealed at 700 °C (Figure 4E–H) demonstrated sharper diffraction rings compared to 600 °C annealing. Ring intensity increased for the HA sample with no new observed peaks (Figure 4E). The SiHA1 sample (Figure 4F) detailed the presence of additional rings related to a combination of HA and rutile, which are shown in (Figure 5). For the higher silicon content coatings, (SiHA2 and SiHA3), the number of rings decreases, reverting back to a HA system, however, it may be seen that in SiHA2 (Figure 4G) some rings relating to rutile remain. This was also the case for the SiHA3 (Figure 4H) sample, however, the rings were more defuse.

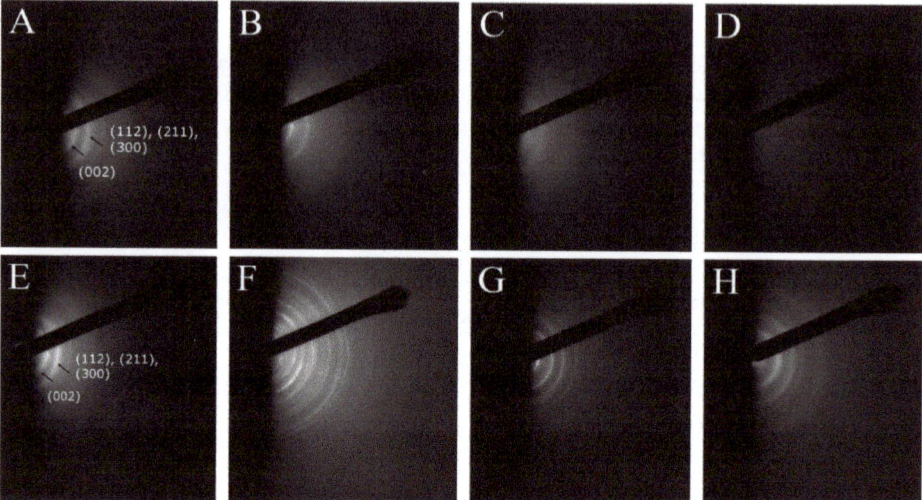

Figure 4. RHEED diffraction patterns of HA and SiHA thin films sputtered onto CPTi discs and annealed at 600 °C and 700 °C for A-D and E-H, respectively. Images were obtained at 200 keV. (**A**) and (**E**) HA, (**B**) and (**F**) SiHA1, (**C**) and (**G**) SiHA2 and (**D**) and (**H**) SiHA3.

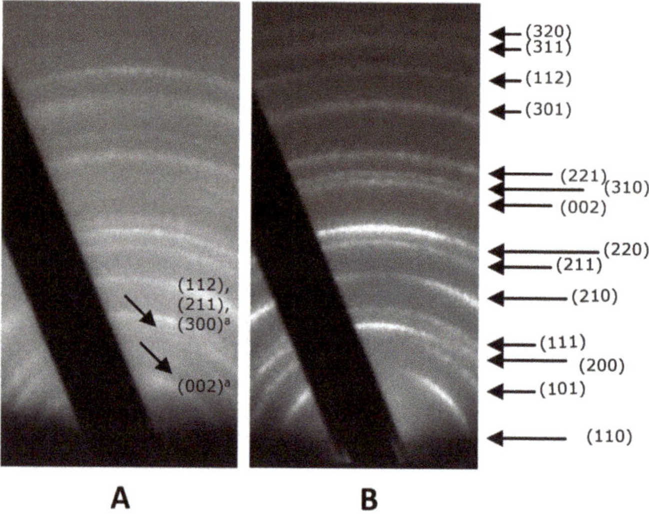

Figure 5. Comparisons of RHEED patterns for (**A**) SiHA1 film on a CPTi disc annealed at 700 °C in flowing argon for 2 h and (**B**) CPTi sample annealed in air at 750 °C for 1 h. All indexed planes match to ICDD card 76-1939 (Rutile) unless indexed with a superscript a plus diagonal arrows indicating possible HA reflections matching to ICDD card 09-432 (HA).

3.1.5. X-Ray Photoelectron Spectroscopy (XPS)

Ca 2p, P 2p and O 1s high resolution XPS spectra are shown in Figure 6. A calcium doublet was observed separated by 3.55 eV and fitted with two components at peak positions of 347.5 and 351.0 eV for Ca 2p1/2 and Ca 2p3/2 respectively [30]. Calcium was in low concentrations in the as deposited films (4.3–6.8 at.%), but increased after annealing at both temperatures (12.1–18.0 at.%). P 2p peaks were fitted with a doublet [31], with separation energy of 0.84 eV. Phosphorus content decreased with both increasing annealing temperature and silicon content. Interestingly, no phosphorus was seen on any of the HA thin films annealed at 600 °C, but was seen on HA samples annealed at 700 °C.

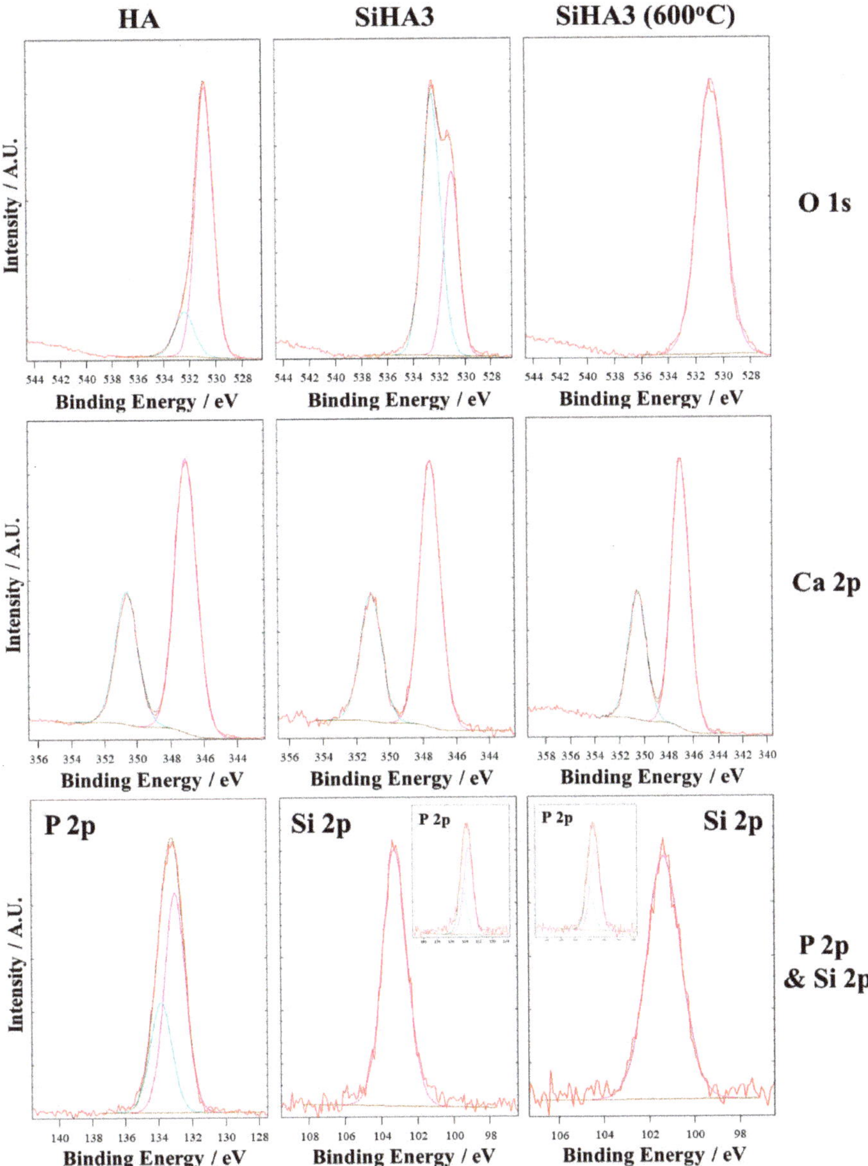

Figure 6. Representative high-resolution XPS spectra for HA, SiHA3 and 600 °C annealed SiHA3 samples, demonstrating the binding energy data for O 1s, Ca 2p, P 2p and Si 2p spectra, where appropriate.

The O 1s peak for the as deposited samples was fitted with two components at 531.0 eV and 532.5 eV, corresponding to PO_4 [32] groups and C–O [32] or SiO_2 groups [33], respectively. The second C–O component became larger with increasing silicon content. Therefore, it is thought that this is related to SiO_2 binding [33]. After annealing at 600 or 700 °C, the O 1s peak could only be fitted to a single component at 531.0 eV, which corresponded to PO_4 bonds. Furthermore, a reduction in oxygen content was seen in all films after annealing, however, no differences were observed between the oxygen content of films annealed at 600 or 700 °C. In the HA samples annealed at 600 °C (which showed no phosphorus present on the surface), no shift in the binding energy of the component at 531 eV was seen.

The Si 2p silicon peak was fitted to a single component. As already shown by the EDX data, the silicon content of thin films increased with increasing power density applied to the silicon target. These values, however, were in poor agreement with the EDX being consistently lower. In the as deposited samples the chemical shift for the Si 2p were found to depend on the silicon content of the film, with lower binding energies measured for the samples with lower silicon concentrations. After annealing at both temperatures all Si 2p peak positions were in the region of 101.5. Silicon content did not vary after annealing at 600 °C, compared with the as deposited samples, however after a heat treatment of 700 °C, only very small quantities were seen on all of the SiHA samples.

The Ca/P ratio decreased with increasing silicon content from 1.43 in the HA samples to 1.03 in the SiHA3 samples for the as deposited samples. After heat treatments of 600 °C the Ca/P ratio increased. This increase was higher for higher concentration silicon containing HA films. Following annealing at 700 °C a further increase in Ca/P ratio was seen.

3.1.6. Fourier Transform Infrared Spectroscopy (FTIR)

FTIR was used to assess the chemical bonding in RF magnetron sputtered thin films. Figure 7A shows infrared spectra for as deposited HA and SiHA thin films sputtered onto CPTi. The HA, SiHA1 and SiHA2 films exhibited four distinct bands at wavenumbers 1147, 1028, 950 and 617 cm^{-1}, which are indicative of υ_3 P–O stretching. SiHA3 samples showed a reduction in the number of phosphate bands, with only peaks at 1147, 950 and 617 cm^{-1} present.

After heat treatments at 600 °C (Figure 7B), the HA films showed sharper peaks with an additional phosphate band at 1080 cm^{-1} when compared to the as deposited HA film. Moreover, a small OH peak was seen at 3643 cm^{-1}. The SiHA1 sample showed all phosphate bonds exhibited by the recrystallised HA sample, however, bands were slightly broader, with a new peak at 820 cm^{-1}. The SiHA2 sample showed broader phosphate bands, and the intensity of the peak at 820 cm^{-1} was reduced. The SiHA3 sample only showed three broad phosphate bands at 1147, 950 and 617 cm^{-1}. This spectrum was very similar to the spectrum of the as deposited coating. Due to similarity of the produced spectra, the 700 °C heat treatment is not shown.

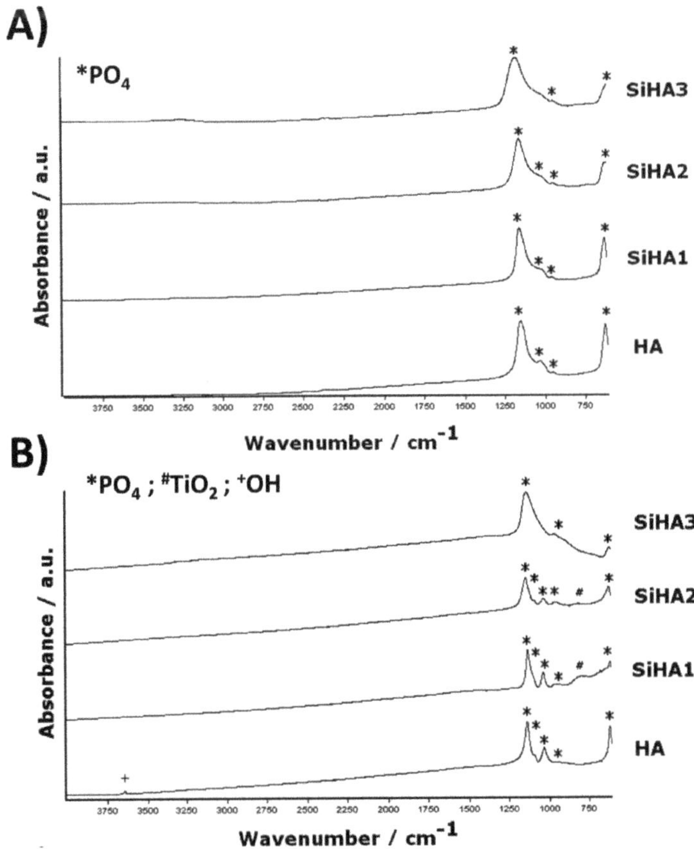

Figure 7. FTIR Spectra of (**A**) as deposited HA and SiHA thin films on CPTi substrates and (**B**) HA and SiHA thin films on CPTi substrates heat treated at 600 °C for 2 h in argon.

3.1.7. Surface Roughness

Stylus profilometry was performed on coatings to assess surface roughness and morphology. Figure 8 shows measured R_a values for uncoated polished CPTi, HA and SiHA coated discs. Uncoated CPTi discs had an R_a value of ca. 16 nm and their roughness increased to ca. 25 and 70 nm after heat treatment at 600 and 700 °C, respectively. As deposited HA and all SiHA coatings exhibited similar roughness, ca. 20 nm. After heat treatments at 600 °C, HA films had the highest roughness value at ca. 41 nm, which decreased gradually with increasing silicon content. This effect was also observed for all films heat treated at 700 °C, but at significantly higher roughness values than the samples treated at 600 °C, with HA films having a roughness of ca. 78 nm, and the SiHA3 samples being measured at ca. 32 nm.

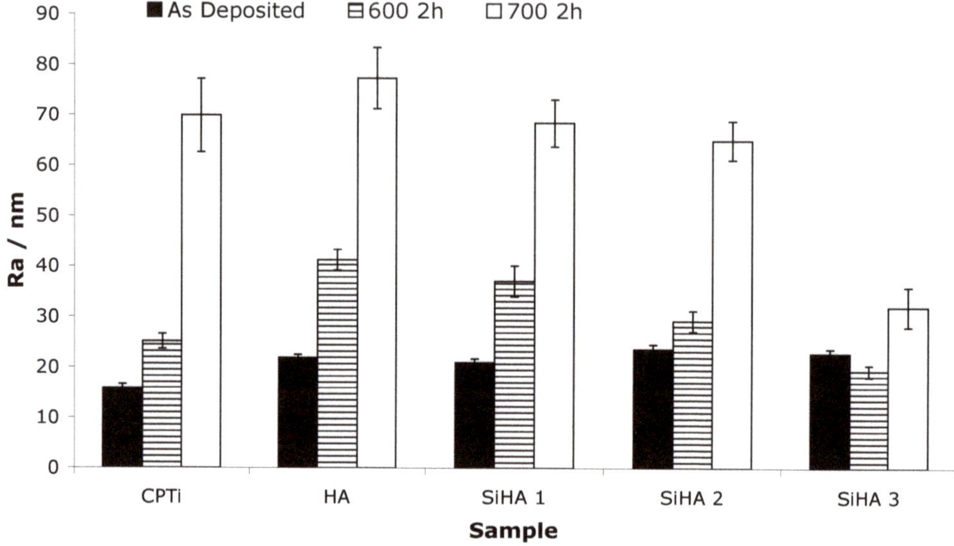

Figure 8. R_a roughness values for CPTi, HA and SiHA thin films as received and heat treated at 600 and 700 °C in flowing argon for 2 h.

3.1.8. Film Wettability

Water contact angle testing was performed to establish the surface hydrophobicity or hydrophilicity of thin film samples. Figure 9 shows measured contact angles for HA and SiHA thin films. All silicon doped samples exhibited lower contact angles than the pure HA films whether in an as deposited or annealed state. The as deposited HA film had a water contact angle of 67°, which decreased with increasing silicon content down to an angle of 27° for the SiHA3 samples. Following heat treatments at 600 °C, the contact angle for all samples decreased when compared to the as deposited samples to values of 54, 41, 31 and 26° for the HA, SiHA1, SiHA2 and SiHA3, respectively. After heat treatments at 700 °C, an increase in contact angle was seen for all samples when compared to either the as deposited or samples heat treated at 600 °C, measuring 69, 56, 42 and 36° for the HA to the SiHA3 samples. Figure 9 also shows optical images of water droplets on as deposited HA and SiHA surfaces. It can therefore be seen that the hydrophilicity increases with increasing silicon content.

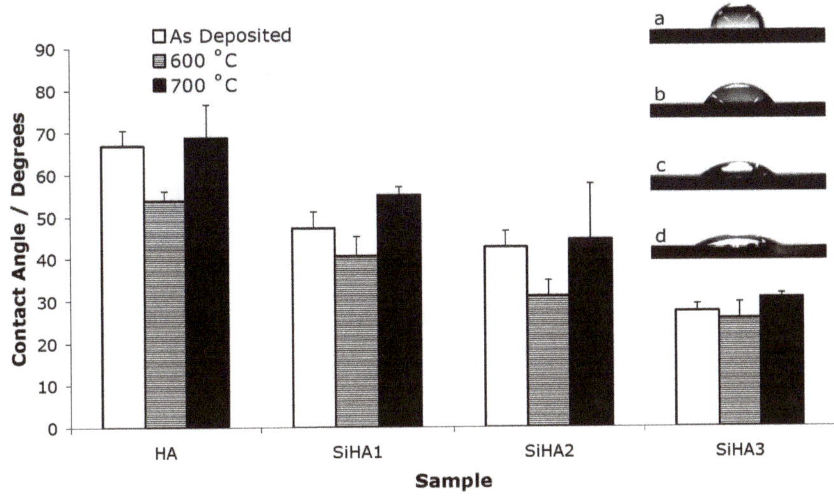

Figure 9. Measured contact angles of water droplets on HA and SiHA thin sputtered films. Mean ± standard error of the mean where n ≥ 6. Also shown are digitally enhanced representative grayscale photographs of water droplets on (**A**) HA (**B**) SiHA1 (**C**) SiHA2 and (**D**) SiHA3 thin sputtered films, showing the effect of silicon doping on the contact angle of water. Photographs taken from as deposited sample set.

3.2. *In Vitro Cytocompatibility Testing*

3.2.1. Elusion Testing – Metabolic Activity, DNA Content and Morphology

Figure 10A_1 shows no significant difference ($p > 0.05$) between the metabolic activity of either the control thermanox samples or the samples in HA and SiHA dissolution media. This was confirmed by the Hoest DNA staining assay (Figure 10A_2), in which there was no significant difference ($p > 0.05$) in DNA content of HOBs grown in the dissolution products of any of the coatings.

Cell morphologies for the control (Figure 10B insert) and all samples (Figure 10B–D) appeared to be similar, showing a monolayer over the thermanox surface. Cells were well spread, appearing to cover similar cell areas. Filopodia and lamellapodia were also observed, indicating cell signalling was occurring successfully and did not demonstrate significant cytotoxicity.

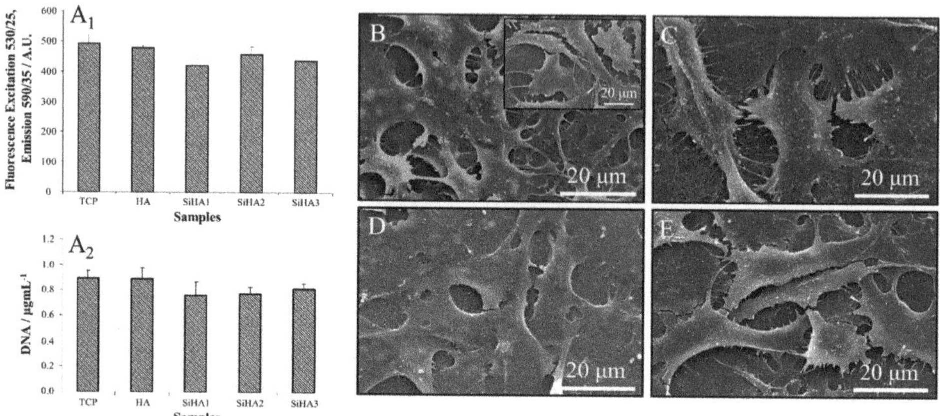

Figure 10. Combined elusion testing results showing: (**A_1**) Metabolic activity and (**A_2**) DNA content of pre-seeded HOB cells exposed to dissolution media of HA and SiHA samples for 24 h (Both expressed as mean ± standard error; n = 6); (**B–D**) SEM micrographs of HOB cell morphology on thermanox slides after 24 h of culture in media containing the dissolution products of (**B**) HA, (**C**) SiHA1, (**D**) SiHA2 and (**E**) SiHA3 thin films (insert image in B shows cells culture in fresh media for reference).

3.2.2. Initial Attachment

Figure 11 shows the adhesion of HOB cells as a percentage of a TCP control. As deposited and heat treated HA thin films showed good attachment after 90 min with values of 51 and 91% of the control, and were found to be significantly different ($p < 0.05$). As deposited and heat treated SiHA1 and SiHA2 samples showed poor cell attachment which was less than 20% of the control. All values for SiHA1 and SiHA2 samples were significantly lower than the heat treated HA thin film ($p < 0.05$). The as deposited SiHA3 samples showed good cell attachment compared to the as deposited and annealed SiHA1 and SiHA2 samples, and were not significantly different from the heat treated HA samples ($p < 0.05$). Following heat treatments at 600 °C for 2 h, the SiHA3 sample showed poor cell adhesion; 7% of the attachment seen on the control surface.

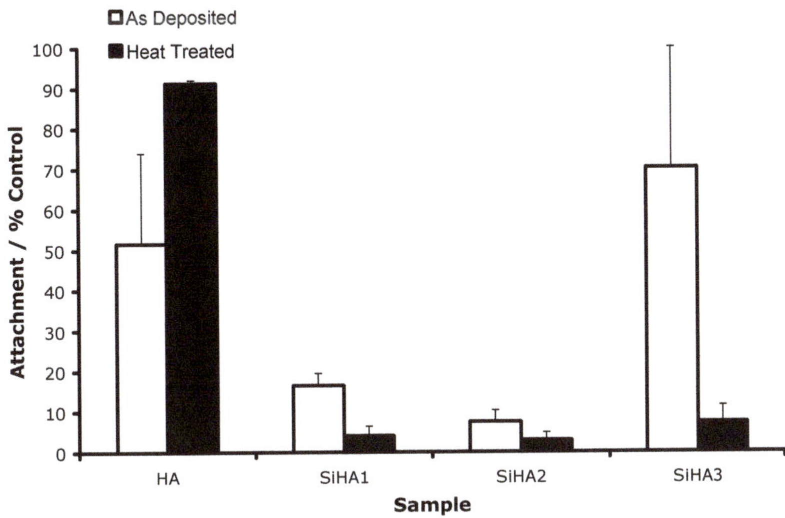

Figure 11. The 90 min attachment of HOB cells to as deposited and heat treated HA and SiHA films at 600 °C. Values are mean ± standard error where n = 6.

3.2.3. Proliferation and Differentiation—Cellular Activity, ALP, and Morphology

The AlamarBlue™ assay results of all sample types annealed at 600 °C (Figure 12A) demonstrate that the HA surface exhibited the highest cellular metabolic activity at every time point. The lowest activity was recorded on the SiHA1 samples. There appeared to be no increase in metabolic activity until day 14, where cell activity significantly increased. Conversely, metabolic activity increased with increasing silicon content of SiHA samples. The SiHA2 and SiHA3 samples showed higher cellular activity compared to the SiHA1 samples, and their metabolic activity increased gradually over time. However, HA and SiHA1 surfaces annealed at 700 °C (Figure 12B) demonstrated similar metabolic activity at all time points, but the differences were not statistically significant ($p > 0.05$). Overall proliferation of both samples was seen to increase gradually up to the 10 d time point, followed by a slight reduction at day 14.

DNA measurements (Figure 12C,D) confirmed the trends shown by the Alamar Blue™ assay, with HA annealed at 600 °C (Figure 12C) exhibiting the highest DNA content, followed by the SiHA3, SiHA2 and then the lowest, exhibited by the SiHA1 samples. HA consistently showed higher DNA values than all other samples. Annealing at 700 °C showed no significant difference between the HA and SiHA1 samples at any time point (Figure 12D).

ALP activity was negligible for all 600 °C annealed surfaces after seven days of culture (Figure 12E). After a time period of 10 days significant ALP activity was seen for the HA samples at approximately 35% of the TCP control. Negligible values were recorded for all silicon containing coatings. For the 700 °C annealed samples (Figure 12F), ALP was not expressed until day 10 for the HA sample, recording a value at 45% of the control with no significant difference at day 14. The SiHA1 samples arguably exhibited ALP production at day seven at approximately 10% of control. This was seen to significantly increase ($p < 0.05$) at day 10 to a value of 50%, which increased further to approximately 55% of the control at day 14.

The morphology of cells on thermanox slides (Figure 12G), HA (Figure 12H) and SiHA3 (Figure 12I) samples annealed at 600 °C at day 14 are shown in Figure 12G–I. SiHA3 was used as a representative for SiHA samples. HOBs on all surfaces appeared to be multi-layered, indicating desirable osteoblast cell growth. Both HA (Figure 12J) and SiHA1 (Figure 12M) surfaces (one day incubation) annealed at 700 °C appeared to be more textured than samples treated at 600 °C. Cells appeared to react to this topography in the case of both samples by larger numbers of extending extra cellular processes compared to the 600 °C sample. After seven days of cell culture (Figure 12K,N), cells on both samples covered the sample surfaces and multilayering had occurred, and no differences were seen between the two samples. Cracks were also seen in both cell samples, which was due to the dehydration protocol adopted. Fourteen-day samples (Figure 12L,O) exhibited cracks again, but no difference in morphologies was seen. Some directional growth can be seen in both samples.

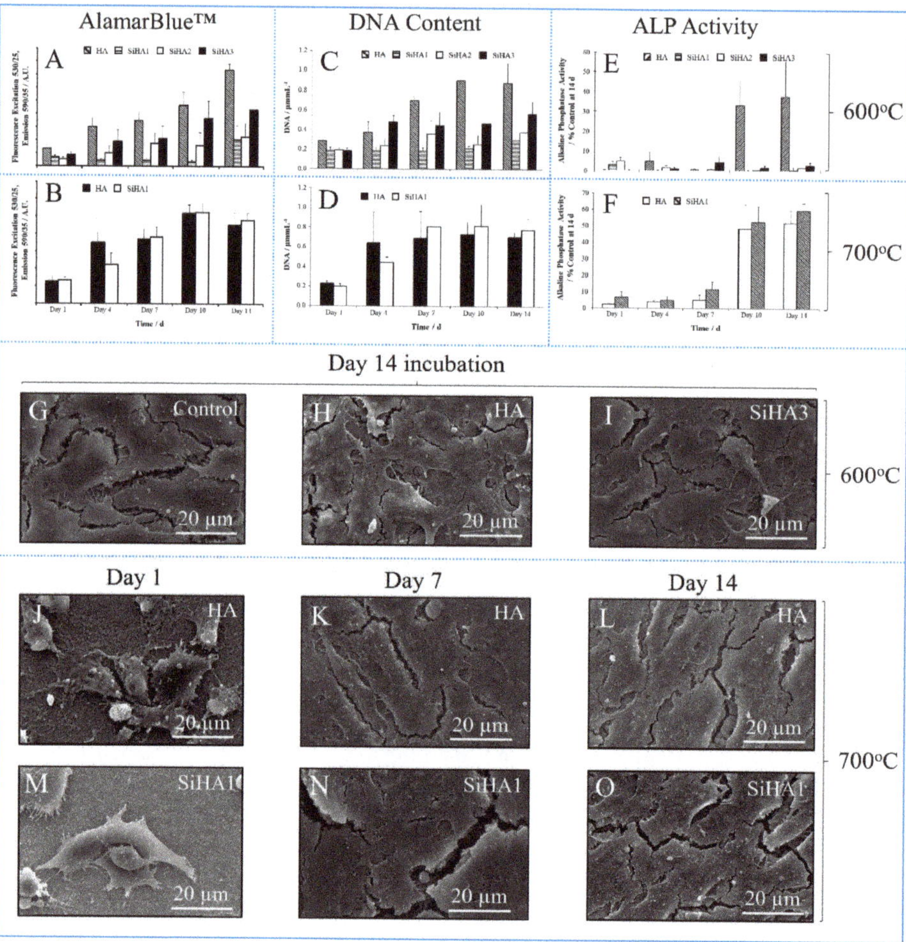

Figure 12. Combined cellular proliferation data showing: AlamarBlue™ assay of HOBs on HA and SiHA surfaces heat treated at (**A**) 600 °C and (**B**) 700 °C; DNA content of HOBs on HA and SiHA surfaces heat treated at (**C**) 600 °C and (**D**) 700 °C; ALP activity of HOBs on HA and SiHA surfaces heat treated at (**E**) 600 °C and (**F**) 700 °C (All graphs are plotted with mean ± standard error of the mean where n = 6); SEM micrographs of cellular morphology showing 14 day incubation on 600 °C annealed (**G**) Thermanox (Control), (**H**) HA, and (**I**) SiHA3; Further SEM micrographs of cell morphology on 700 °C annealed HA and SiHA1 samples incubated at day (**J**),(**M**) 1, (**K**),(**N**) 7 and (**L**),(**O**) 14, respectively.

4. Discussion

4.1. Composition and Topographical Analysis

Studies on bulk SiHA, as well as thick and thin film SiHA coatings, have identified them as eliciting an enhanced cellular and tissue response compared to HA [22,28,34–36]. The characterisation of thin film materials is still limited, despite detailed articles within the literature [25,37–45]. As a result, the role of silicon in the HA crystal structure is still not fully understood in the metastable system descried here and, moreover, how this interacts with cells; hence the importance and role of this study.

Morphologically, the surface (Figure 2) and cross-sectional (Figure 1) results, in conjunction with the XRD (Figure 3) and RHEED (Figures 4 and 5) analysis demonstrated clear trends regarding crystallisation of the films, in addition to increasing silicon content, which is in good agreement with Agyapong et al. [29] and Wang et al. [46] The films showed an increase in thickness with increasing power density applied to the Si target confirmed via TEM (Figure 1), which is to be expected since increasing the power density causes an increase in sputtering yield [47]. Furthermore, as is expected with magnetron sputtering deposition [30], the film exhibited good step coverage, as seen in the SEM (Figure 2) and roughness measurements (Figure 8). Silicon content had no effect on deposited film roughness but an increase in roughness was seen after annealing at 600 °C, except the SiHA3 sample, which remained constant. At higher annealing temperatures (700 °C), roughness typically doubled on all sample types; as crystals grow an increase in surface roughnesses will be observed. Furthermore, the higher the recrystallisation temperature the larger the crystallite size (Table 1) and in turn the more textured the surface (Figure 2). However, silicon inclusion has been shown by a number of authors to inhibit the crystallite growth of HA in both bulk and coatings [34], corroborating the current data, where roughness increases by a smaller amount for each increase in silicon addition to the point where the 13.4 wt.% SiHA shows no change in surface roughness (Figure 8).

For SiHA materials, it has been shown that structural configuration is important when trying to enhance osteoblast response [48]. Silicate (SiO_4^{4-}) is considered soluble, whereas silica (SiO_2) is insoluble in water, with Balas et al. [48] identifying that SiO_4^{4-}, due to enhanced solubility, generated a more favourable cellular response. XPS (Figure 6) found that binding energy shifts were consistent with increasing silicon content, with Okada et al. [49] and Stevenson et al. [50] confirming this trend. If fewer silicon atoms are present on the surface it is more probable that silicon will bond to oxygen, with higher silicon concentrations causing polymerisation; sharing electrons between silicon atoms. The increase in binding energy signifies a chemical change in Si–O bonding from polymeric SiO_2 (Q^4) to monomeric SiO_4^{4-} (Q^0) structure; Q^n where n represents the number of bridging oxygen atoms per SiO_4 tetrahedron. Post annealing (600/700 °C) exhibited Si 2p 3/2 binding energies of approximately 101.5 eV, suggesting a Q^0 structure, as detailed in two independent studies [48,51]; SiO_4^{4-} had successfully substituted for PO_4^{3-} groups. However, Balas et al. [48] demonstrated that this effect only occurred with up to 1.6 wt.% silicon addition in bulk materials, above which, it reverted back to a Q^4 configuration (103 eV). In the current study however, up to 13.4 wt.% (bulk)/6 at.% (surface) was successfully substituted into the HA films; a 'super saturated' state compared to theoretical values of 5 wt.% being substituted for PO_4^{3-} tetrahedra in bulk SiHA [52]. The position of the silicate tetrahedra may occupy PO_4^{3-} vacancies, however, further proof is required since OH site doping, leading to $Ca_{10}(PO_4)_4(SiO_4)_2$, could be possible, as no OH groups are seen in SiHA samples from FTIR analysis (Figure 7). However, the above phase would be detected in the XRD (Figure 3) and FTIR (Figure 7) data, but this was not the case, giving further evidence of a silicate substituted hydroxyapatite structure.

Ca/P ratios were significantly different from stoichiometric HA, with clear discrepancies between the EDX (bulk) and XPS (surface) data (Table 1), with EDX determining Ca/P ratios varied from 1.68 to 1.80, potentially from CaO formation [53]. Ratios obtained from XPS were all lower than EDX values (Table 1), likely due to preferential phosphate sputtering [54]. The Ca/P ratios of the as deposited samples decreased with increasing silicon from 1.43 to 1.03, potentially through the sputtering environment allowing the formation of P–Si–O bonds. This is a likely scenario as a large number of Ca–P, Ca–O and P–O like species have been found in HA plasmas [53]. As more silicon is made available due to increasing bias on the silicon target, more P–Si–O groups may form, thus lowering the Ca/P ratio. Furthermore, it was seen that on annealing, the Ca/P ratio increased due to a loss of phosphorus [55,56].

From the FTIR data (Figure 7), increasing the silicon content caused widening of the PO_4 bands, suggesting bond formation is inhibited through silicon addition, even on annealed samples. The presence of OH bonding in the HA sample, which disappeared in the SiHA samples, is due to SiO_4^{4-} species substituting for PO_4^{3-} bonds; an imbalance of −1 is created. The most energetically favourable method of reducing this effect is to reduce the number of OH groups associated with the molecule [34,57]. The numbers of substitutions will be indirectly proportional to the number of OH groups. XPS (Figure 6) confirmed the super saturated state, which may explain why no OH was seen, opposed to just a reduction in OH peak intensity. This effect has also been shown in other apatite systems [58].

Wettability testing (Figure 9) showed that the incorporation of silicon into the HA lattice led to a more hydrophilic surface [51,59], demonstrating that SiHA has a more negative surface charge and increased surface adhesion than HA. Takeda and Fukawa [60] found that OH groups were a major factor governing surface chemical properties of oxide thin films. Higher contact angles were obtained for as deposited samples than samples annealed at 600 °C. When the samples were annealed, residual stresses could be corrected for, thus lowering the surface energy. This was confirmed by the small difference seen between the measured contact angles of as deposited and 600 °C SiHA3 samples, as this sample did not recrystallise at this temperature. After annealing at 700 °C, contact angles were higher than both the as deposited and 600 °C samples, likely due to the appearance of titanium and rutile phases at the sample's surface, as demonstrated from the RHEED analysis (Figures 4 and 5) [61].

As deposited HA and SiHA thin films were shown to be amorphous, with all samples except the SiHA3 sample showing a single phase HA structure post annealing (Figure 3); the SiHA3 samples required 800 °C annealing to recrystallise. Gibson et al. [34] and others have shown that introducing silicon into HA lowers the thermal stability. It has commonly been shown that silicon additions of 5 wt.% or more causes HA, on sintering, to decompose into undesirable phases like CaO and α- or β-TCP. However, no secondary phases were found in any of the films at any annealing temperature. This evidence, in conjunction with the XPS data (Figure 6), further suggests that higher amount of silicon may be substituted in the HA thin film structure than previously reported elsewhere. Crystallinity decreased with increasing silicon content, as confirmed by Zou et al. [43], for lower Si contents (0.8–2.0 wt.%). After annealing at 700 °C the SiHA1 samples showed rutile diffraction patterns, also confirmed via XPS (Figure 6) and RHEED (Figures 4 and 5).

4.2. *In Vitro Cytocompatibility*

Initial cell adhesion studies carried out on as deposited and HA thin films annealed at 600 °C demonstrated cells preferentially adhered to HA surfaces, with poor adhesion on all SiHA surfaces, with the exception of the SiHA3 as deposited samples, which showed good adhesion (Figure 11). Furthermore, the 14 day cell assays using HA and SiHA samples annealed at 600 °C provided further evidence that HOB cells preferred HA to the SiHA surfaces (Figure 12). Osteoblasts on HA surfaces showed increased proliferation but also were seen to be differentiating, indicated by the ALP activity. This affect was not seen on any of the SiHA surfaces after 14 days. This result was considered surprising, as a large amount of literature has been published demonstrating that SiHA ceramics lead to increased proliferation and differentiating activity of osteoblast cells with both bulk and coating materials in vitro and in vivo [62,63]. Assays ruled out that the samples had toxic effects on cells, moreover studies have shown that increased quantities of silicon in cell media can lead to the up-regulation of genes that aid cell proliferation and differentiation [64,65]. It is likely no enhancing effect was seen in our elution study because the test only exposed osteoblasts to media for one day (Figure 10). The stability of SiHA thin films in solution must be considered to be responsible for the poor adhesion and therefore the low proliferation compared with HA films. Qualitative EDX of samples used for the cytotoxicity study (see

Figure A1) confirmed that SiHA surfaces annealed at 600 °C are unstable; dissolving in cell culture media within hours for the highest doped sample. Cells have been shown to attach poorly to highly soluble (bioactive) surfaces, with the converse being noted for stable surfaces [66]. Despite cellular preference for HA surfaces, it was shown that cell number occurred in the order of SiHA3 > SiHA2 > SiHA1, from highest to lowest. When samples are immersed in cell culture media, serum proteins will attach to the surface allowing subsequent attachment of osteoblast cells. Over time, the film will dissolve, taking away with it attached adhesion proteins. Proteins will then change confirmation not allowing cells to attach. A new conditioning layer of protein will then redeposit but will again be removed by dissolution. If cells do manage to attach, they will subsequently be removed with the protein layer. When the protein changes confirmation, the cell will no longer be able to adhere and so will be released. This process will happen continually until cells can attach to a stable surface such as the underlying CPTi substrate. In the case of the SiHA1 compositions, this event does not occur until after day 10, but the dissolution rate was high enough to inhibit long term adhesion. SiHA2 however, showed increasing proliferation with time demonstrated by both the Alamarblue™ and the DNA assay (Figure 12). Again, from XPS data (not shown), only 1–2 at.% of the SiHA3 thin film remained after 2 days in cell culture media. Furthermore, it was observed in the contact angle testing experiments that some film dissolution would occur even when exposed to water for a few minutes. These observations and measurements may explain why initial adhesion of HOB cells is possible and sustainable on the as deposited and 600 °C SiHA3 films (Figure 11). Essentially, the CPTi substrates are revealed to cells which act as a stable protein mediated adhesion site. Initial adhesion studies comparing titanium and HA surfaces have demonstrated that titanium surfaces show a better response in a 90 min attachment period [67], but this was not seen in the case of the as deposited SiHA3 samples, which is thought to occur due to some cells undergoing apoptosis or programmed cell death during the prolonged attachment time. It is well known that cell adhesion via proteins allows signalling which can inhibit apoptosis [68]. Cell adhesion to the substrate via proteins is also necessary for a musculoskeletal cell's vitality, growth, migration, and differentiation [69,70].

It has been shown both in this study and other studies that SiHA bulk and thin film materials have a higher dissolution potential than HA [24,71]. Moreover, the staining protocol required for the initial adhesion may further affect the stability of the surface owing to numerous washing steps involved, accelerating film dissolution and removal of any adhered cells. In the current study, we investigated silicon contents higher than previously reported, ranging from 1.8–13.4 wt.%. Furthermore, coating thicknesses were higher, which tends to lead to higher residual stress in the films and on recrystallisation will give a higher crystallinity. In comparison to bulk materials, Arcos et al. [52] investigated the in vitro response of osteoblast cells to bulk high quantity silicon doped apatites. It was found that high silicon content apatite (low crystallinity) showed poor cell proliferation over a seven day period. This was explained by cells poorly adapting to their environments, however, it is more likely that this is due to surface dissolution inhibiting cell adhesion.

In order to overcome high dissolution rates of samples annealed at 600 °C, the cellular response of samples annealed at 700 °C were investigated, however, due to the reduced crystallinity and lower stability of the SiHA2 and SiHA3 samples, only the SiHA1 sample was investigated. Proliferation and differentiation on SiHA1 surfaces were slightly higher than on HA surfaces, however, this was not significantly different ($p > 0.05$). This conflicts with a large number of studies providing strong evidence that SiHA materials elicit an enhanced response when compared with HA materials [22–24,27,28,35,39,40,71–75]. This may be explained by the presence of a HA/rutile phases at the surface of the 700 °C samples. Moreover, a much lower silicon content was seen on the surface of the samples annealed at 700 °C compared to that of those annealed at 600 °C. Even so, it has been shown that even 0.4 wt.% silicon addition to HA can have a pronounced effect on adhesion and proliferation [27]. As almost no silicon content is present on the SiHA1 after annealing at 700 °C, it would be expected that this surface would have the lowest dissolution rate and be unlikely to cause problems for cell adhesion, but it may in fact have a beneficial effect leading to the slightly increased cellular response, although this was not shown to be significantly different ($p > 0.05$) from the HA sample.

Commonly in the literature, it has been seen that increasing the post deposition temperature of HA ceramics increases the cell proliferation and differentiation in both bulk and thin film systems [67,76,77]. Roughness, topography chemistry and surface energy are all known to influence cell response to a given surface [78]. Data obtained would suggest that in the current study cells have reacted to the roughness and chemistry. The majority of studies concerned with topography have concentrated on the micron scale, with only a few authors concentrating on the nanometre scale. This is mainly due to a lack of knowledge of how to produce such surfaces, however Affrossman et al. [79] have used polymer demixing to achieve nanotopographies. It has been reported that cells can detect changes as small as 5 nm and in vivo cells commonly respond to 66 nm banding on collagen fibrils [80]. In this study there was a roughness difference of 35 nm and cells were shown to react to this will increasing numbers of lamellapodia and filopodia leading to distinct attachment sites. This led to no significant difference ($p > 0.05$) in cell number and metabolic activity, suggesting that roughness values on this scale have no major effects on cellular response. Recently, Kahn et al. [81] used neural cells to investigate several surface textures ranging from 10 to 250 nm in roughness. Values between 20–100 nm promoted cell adhesion and longevity, however, surfaces led to a decrease in attachment at values > 100 nm. Similar trends have been found in other studies using different cells [82], but it is often the case that differences as low as 30 nm did not yield any notable difference. Dalby et al. [83] studied the effect of nano-islands on polystyrene materials with fibroblasts. It was found that islands as low as 13 nm high led to increased adhesion, proliferation and cytoskeletal development when compared to flat controls. Conversely, nano-islands 95 nm in height lead to unusual, stellate morphologies with poorly formed cytoskeletons [84]. Intermediate islands (45 nm) showed no difference in cell area from the control, however the cytoskeleton was less well formed. Studies have shown that RF magnetron sputtered HA surfaces show no significant difference ($p > 0.05$), when compared with titanium substrates at initial time points [28,85] and the current study agrees with such work. It may, however, be that because phosphorus was not found at the top few atomic layers the cellular response was impaired. While not directly comparable, it has been shown that cells respond preferentially to surfaces with stoichiometric Ca/P values [86]. The literature confirms that surface texture and chemistry are important, but it is still under debate which has a more positive effect.

Overall, the combination of nanotopography and change in surface chemistry has led to small changes in cell morphology and proliferation over a 14 day time period, however such differences in the HA and SiHA1 surfaces annealed at 700 °C for 2 h were too subtle to be significantly different.

5. Conclusions

The work performed in this study investigated HA and SiHA RF/Pulsed DC magnetron sputtered thin films as coatings for orthopaedic applications. As deposited HA thin films were found to be amorphous or nanocrystalline, however, upon annealing (600 °C for all samples, except SiHA3, which required 800 °C) recrystallised. Furthermore, the addition of silicon to HA thin films inhibited HA crystallite growth as demonstrated by crystallite sizes calculated from XRD line broadening.

Both EDX and XPS showed a reduction in Ca/P ratio with increasing silicon content due to possible creation of a P–Si–O chemical species during deposition, which allows P to reach the substrate more readily. After annealing, however, the Ca/P ratio increased with increasing temperature, likely due to the evaporation of volatile phosphate species, facilitated by silicon inclusion destabilising the HA structure. XPS demonstrated that deposited SiHA thin films contain polymerised silicate networks, transforming to monomeric states after annealing, suggesting $SiO_4{}^{4-}$ substitution in the HA lattice for $PO_4{}^{3-}$ chemical species.

The roughness was shown to be similar for all as deposited films, which were measured to be approximately 20 nm (Ra), similar to the CP-Ti substrate, which increased following annealing; this was inversely proportional to the silicon content, due to silicon inhibition of the HA crystallite growth and the rise of rutile grain growth. Silicon is also known to inhibit rutile growth, thus explaining the lowering in roughness with increasing silicon content. Water contact angle testing demonstrated silicon addition to the HA structure increased hydrophilicity with increasing silicon content.

Initial adhesion, proliferation and differentiation assays all suggested HOBs preferred HA to the SiHA surfaces, due to silicon doping destabilising the HA thin films, removing the protein conditioning layer essential for normal cell adhesion and growth. HOBs on the highest silicon doped HA thin films annealed at 600 °C showed some proliferation due to the stable CPTi substrate surface becoming available for protein mediated cell adhesion. After annealing at 700 °C, no significant difference ($p > 0.05$) was seen between the HA and the SiHA1 surfaces, suggesting enhanced cellular response due to crystallinity levels.

This study ultimately demonstrates that for higher (one of the highest tested in the literature), meta-stable doping levels of Si into the HA structure, cellular response is strongly linked to the crystallinity of the produced HA and SiHA films, the surface stability, as well as other properties, such as surface wettability, roughness, etc. Despite literature studies showing small doping levels of Si have a positive effect on cellular proliferation, this is not seen in higher-doped systems and, therefore, careful optimisation is required to glean appropriate properties.

Author Contributions: Conceptualization, S.C.C., G.S.W., C.A.S. and D.M.G.; methodology, S.C.C., G.S.W., C.A.S. and D.M.G.; validation, S.C.C., M.D.W., R.M.F., I.A., G.S.W., C.A.S. and D.M.G.; formal analysis, S.C.C., and M.D.W.; investigation, S.C.C.; resources, S.C.C., G.S.W., C.A.S. and D.M.G.; data curation, S.C.C. and M.D.W.; writing—original draft preparation, S.C.C., M.D.W., R.M.F., and D.M.G.; writing—review and editing, S.C.C., M.D.W., R.M.F., I.A., G.S.W., C.A.S. and D.M.G.; visualization, S.C.C., M.D.W., R.M.F., I.A., G.S.W., C.A.S. and D.M.G.; supervision, G.S.W., C.A.S. and D.M.G.; project administration, G.S.W., C.A.S. and D.M.G.; funding acquisition, G.S.W., C.A.S. and D.M.G. All authors have read and agreed to the published version of the manuscript.

Funding: This work was supported through funding from Teer Coatings Ltd.

Acknowledgments: The authors would like to thank all technical staff who aided in equipment operation, particularly Keith Dinsdale, George Anderson, Martin Roe, Nigel Neate, Julie Thornhill, Tom Buss, Rory Screaton, and Graham Malkinson.

Conflicts of Interest: The authors declare no conflict of interest.

Appendix A

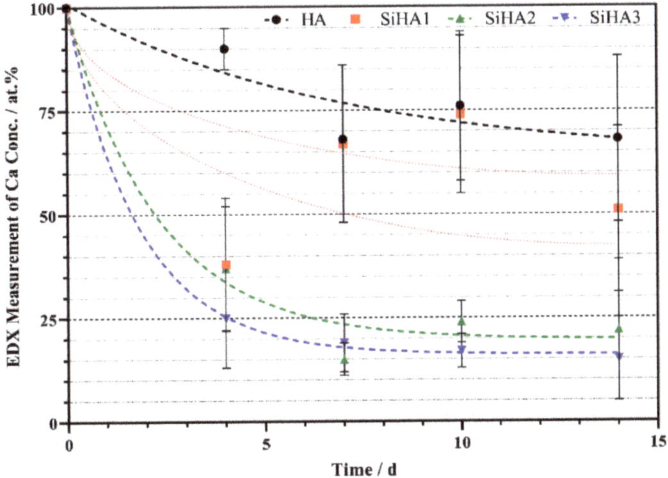

Figure A1. Percentage of HA and SiHA thin films annealed at 600 °C remaining as a percentage of original coating Ca content. Data plotted is mean ± standard error of the mean; n = 4. Non-linear one-phase decay regression plots were calculated using GraphPad Prism software based on the data shown. Due to the high variance of the SiHA1 sample, a suitable regression line was unable to be plotted, hence a probable area has manually been fitted for visual enhancement.

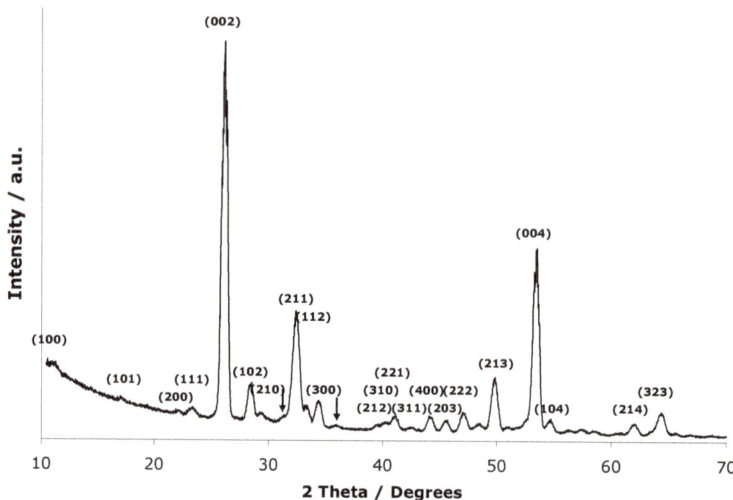

Figure A2. XRD plot for a Plasma Biotal plasma sprayed copper backed target. All major diffraction planes are indexed. Arrows indicate potential β-TCP secondary phase, however, with the low intensity accurate quantification is difficult.

References

1. Cao, W.P.; Hench, L.L. Bioactive materials. *Ceram. Int.* **1996**, *22*, 493–507. [CrossRef]
2. Narayanan, R.; Seshadri, S.K.; Kwon, T.Y.; Kim, K.H. Calcium phosphate-based coatings on titanium and its alloys. *J. Biomed. Mater. Res. B Appl. Biomater.* **2008**, *85*, 279–299. [CrossRef] [PubMed]
3. Stuart, B.W.; Gimeno-Fabra, M.; Segal, J.; Ahmed, I.; Grant, D.M. Mechanical, structural and dissolution properties of heat treated thin-film phosphate based glasses. *Appl. Surf. Sci.* **2017**, *416*, 605–617. [CrossRef]
4. Schrooten, J.; Helsen, J.A. Adhesion of bioactive glass coating to Ti6Al4V oral implant. *Biomaterials* **2000**, *21*, 1461–1469. [CrossRef]
5. Chung, C.J.; Long, H.Y. Systematic strontium substitution in hydroxyapatite coatings on titanium via micro-arc treatment and their osteoblast/osteoclast responses. *Acta Biomater.* **2011**, *7*, 4081–4087. [CrossRef]
6. Lim, J.Y.; Donahue, H.J. Cell sensing and response to micro- and nanostructured surfaces produced by chemical and topographic patterning. *Tissue Eng.* **2007**, *13*, 1879–1891. [CrossRef]
7. Suchanek, W.; Yoshimura, M. Processing and properties of hydroxyapatite-based biomaterials for use as hard tissue replacement implants. *J. Mater. Res.* **1998**, *13*, 94–117. [CrossRef]
8. Nawawi, N.; Alqap, A.S.F.; Sopyan, I. Recent progress on hydroxyapatite-based dense biomaterials for load bearing bone substitutes. *Recent Pat. Mater. Sci.* **2011**, *4*, 63–80. [CrossRef]
9. Mohseni, E.; Zalnezhad, E.; Bushroa, A.R. Comparative investigation on the adhesion of hydroxyapatite coating on Ti–6Al–4V implant: A review paper. *Int. J. Adhes. Adhes.* **2014**, *48*, 238–257. [CrossRef]
10. Wadge, M.D.; Stuart, B.W.; Thomas, K.G.; Grant, D.M. Generation and characterisation of gallium titanate surfaces through hydrothermal ion-exchange processes. *Mater. Des.* **2018**, *155*, 264–277. [CrossRef]
11. Sun, L.; Berndt, C.C.; Gross, K.A.; Kucuk, A. Material fundamentals and clinical performance of plasma-sprayed hydroxyapatite coatings: A review. *J. Biomed. Mater. Res.* **2001**, *58*, 570–592. [CrossRef] [PubMed]
12. Cheang, P.; Khor, K.A. Addressing processing problems associated with plasma spraying of hydroxyapatite coatings. *Biomaterials* **1996**, *17*, 537–544. [CrossRef]
13. Yang, Y.-C.; Chang, E. Influence of residual stress on bonding strength and fracture of plasma-sprayed hydroxyapatite coatings on Ti–6Al–4V substrate. *Biomaterials* **2001**, *22*, 1827–1836. [CrossRef]
14. Mattox, D.M. *Handbook of Physical Vapor Deposition (PVD) Processing*; William Andrew: Norwich, NY, USA, 2010.

15. Wadge, M.D.; Turgut, B.; Murray, J.W.; Stuart, B.W.; Felfel, R.M.; Ahmed, I.; Grant, D.M. Developing highly nanoporous titanate structures via wet chemical conversion of DC magnetron sputtered titanium thin films. *J. Colloid Interface Sci.* **2020**, *566*, 271–283. [CrossRef] [PubMed]
16. Pichugin, V.F.; Surmenev, R.A.; Shesterikov, E.V.; Ryabtseva, M.; Eshenko, E.; Tverdokhlebov, S.I.; Prymak, O.; Epple, M. The preparation of calcium phosphate coatings on titanium and nickel–titanium by rf-magnetron-sputtered deposition: Composition, structure and micromechanical properties. *Surf. Coat. Technol.* **2008**, *202*, 3913–3920. [CrossRef]
17. van Dijk, K.; Schaeken, H.G.; Wolke, J.G.; Jansen, J.A. Influence of annealing temperature on RF magnetron sputtered calcium phosphate coatings. *Biomaterials* **1996**, *17*, 405–410. [CrossRef]
18. Yang, Y.; Kim, K.-H.; Ong, J.L. A review on calcium phosphate coatings produced using a sputtering process—an alternative to plasma spraying. *Biomaterials* **2005**, *26*, 327–337. [CrossRef]
19. Jeong, Y.-H.; Choe, H.-C.; Brantley, W.A. Hydroxyapatite-silicon film deposited on Ti–Nb–10Zr by electrochemical and magnetron sputtering method. *Thin Solid Film.* **2016**, *620*, 114–118. [CrossRef]
20. Kozelskaya, A.; Kulkova, S.; Fedotkin, A.; Bolbasov, E.; Zhukov, Y.; Stipniece, L.; Bakulin, A.; Useinov, A.; Shesterikov, E.; Locs, J. Radio frequency magnetron sputtering of Sr-and Mg-substituted β-tricalcium phosphate: Analysis of the physicochemical properties and deposition rate of coatings. *Appl. Surf. Sci.* **2019**, 144763. [CrossRef]
21. Graziani, G.; Bianchi, M.; Sassoni, E.; Russo, A.; Marcacci, M. Ion-substituted calcium phosphate coatings deposited by plasma-assisted techniques: A review. *Mater. Sci. Eng. C Mater. Biol. Appl.* **2017**, *74*, 219–229. [CrossRef]
22. Thian, E.S.; Huang, J.; Best, S.M.; Barber, Z.H.; Bonfield, W. Silicon-substituted hydroxyapatite: The next generation of bioactive coatings. *Mater. Sci. Eng. C* **2007**, *27*, 251–256. [CrossRef]
23. Thian, E.S.; Huang, J.; Vickers, M.E.; Best, S.M.; Barber, Z.H.; Bonfield, W. Silicon-substituted hydroxyapatite (SiHA): A novel calcium phosphate coating for biomedical applications. *J. Mater. Sci.* **2006**, *41*, 709–717. [CrossRef]
24. Thian, E.S.; Huang, J.; Best, S.M.; Barber, Z.H.; Bonfield, W. Novel silicon-doped hydroxyapatite (Si-HA) for biomedical coatings: An in vitro study using acellular simulated body fluid. *J. Biomed. Mater. Res. Part B Appl. Biomater.* **2006**, *76*, 326–333. [CrossRef] [PubMed]
25. Surmeneva, M.A.; Mukhametkaliyev, T.M.; Tyurin, A.I.; Teresov, A.D.; Koval, N.N.; Pirozhkova, T.S.; Shuvarin, I.A.; Shuklinov, A.V.; Zhigachev, A.O.; Oehr, C.; et al. Effect of silicate doping on the structure and mechanical properties of thin nanostructured RF magnetron sputter-deposited hydroxyapatite films. *Surf. Coat. Tech.* **2015**, *275*, 176–184. [CrossRef]
26. Qi, J.; Chen, Z.; Han, W.; He, D.; Yang, Y.; Wang, Q. Effect of deposition parameters and heat-treatment on the microstructure, mechanical and electrochemical properties of hydroxyapatite/titanium coating deposited on Ti6Al4V by RF-magnetron sputtering. *Mater. Res. Express* **2017**, *4*, 096409. [CrossRef]
27. Thian, E.S.; Huang, J.; Best, S.M.; Barber, Z.H.; Bonfield, W. A new way of incorporating silicon in hydroxyapatite (Si-HA) as thin films. *J. Mater. Sci. Mater. Med.* **2005**, *16*, 411. [CrossRef]
28. San Thian, E.; Huang, J.; Best, S.M.; Barber, Z.H.; Brooks, R.A.; Rushton, N.; Bonfield, W. The response of osteoblasts to nanocrystalline silicon-substituted hydroxyapatite thin films. *Biomaterials* **2006**, *27*, 2692–2698. [CrossRef]
29. Agyapong, D.A.Y.; Zeng, H.; Du, W.; Acquah, I.; Liu, Y. Comparative analysis of the structural and physical properties of magnetron Co-sputtered Ag-doped and Si-doped hydroxyapatite coatings on titanium substrates. *Integr. Ferroelectr.* **2017**, *180*, 69–76. [CrossRef]
30. Boyd, A.R.; Meenan, B.J.; Leyland, N.S. Surface characterisation of the evolving nature of radio frequency (RF) magnetron sputter deposited calcium phosphate thin films after exposure to physiological solution. *Surf. Coat. Technol.* **2006**, *200*, 6002. [CrossRef]
31. Boyd, A.; Akay, M.; Meenan, B.J. Influence of target surface degredation on the properties of r.f. magnetron-sputtered calcium phosphate coatings. *Surf. Interface Anal.* **2003**, *35*, 188–198. [CrossRef]
32. Yamaguchi, T.; Tanaka, Y.; Ide-Ektessabi, A. Fabrication of hydroxyapatite thin films for biomedical applications using RF magnetron sputtering. *Nucl. Instrum. Methods Phys. Res. Sect. B Beam Interact. Mater. At.* **2006**, *249*, 723. [CrossRef]
33. Wagner, C.D.; Naumkin, A.V.; Kraut-Vass, A.; Allison, J.W.; Powell, C.J., Jr. *NIST X-Ray Photoelectron Spectroscopy Database*; John Wiley & Sons Ltd.: Hoboken, NJ, USA, 2000.

34. Gibson, I.R.; Best, S.M.; Bonfield, W. Chemical characterization of silicon-substituted hydroxyaptite. *J. Biomed. Mater. Res.* **1999**, *44*, 422–428. [CrossRef]
35. Gibson, I.R.; Hing, K.; Revell, P.A.; Santos, J.D.; Best, S.M.; Bonfield, W. Enhanced in vivo response to silicate-substituted hydroxyapaptite. *Key Eng. Mater.* **2002**, *218-220*, 203–206. [CrossRef]
36. Qadir, M.; Li, Y.; Wen, C. Ion-substituted calcium phosphate coatings by physical vapor deposition magnetron sputtering for biomedical applications: A review. *Acta Biomater.* **2019**, *89*, 14–32. [CrossRef] [PubMed]
37. Thian, S.; Huang, J.; Barber, Z.H.; Best, S.M.; Bonfield, W. Surface modification of magnetron-sputtered hydroxyapatite thin films via silicon substitution for orthopaedic and dental applications. *Surf. Coat. Tech.* **2011**, *205*, 3472–3477. [CrossRef]
38. Thian, E.S.; Haung, J.; Best, S.M.; Barber, Z.H.; Bonfield, W. Surface wettability enhances osteoblast adhesion on silicon-substituted hydroxyapatite thin films. *Key Eng. Mater.* **2007**, *330-332*, 877–880. [CrossRef]
39. Thian, E.S.; Huang, J.; Best, S.M.; Barber, Z.H.; Bonfield, W. Silicon-substituted hydroxyapatite thin films: Effect of annealing temperature on coating stability and bioactivity. *J. Biomed. Mater. Res.* **2006**, *78*, 121–128. [CrossRef]
40. Porter, A.E.; Rea, S.M.; Galtrey, M.; Best, S.M.; Barber, Z.H. Production of thin film silicon-doped hydroxyapatite via sputter deposition. *J. Mater. Sci.* **2004**, *39*, 1895–1898. [CrossRef]
41. Zou, S.; Huang, J.; Best, S.M.; Bonfield, W. Crystal imperfection studies of pure and silicon substituted hydroxyapatite using Raman and XRD. *J. Mater. Sci. Mater. Med.* **2005**, *16*, 1143–1148. [CrossRef]
42. Surmeneva, M.A.; Chaikina, M.V.; Zaikovskiy, V.I.; Pichugin, V.F.; Buck, V.; Prymak, O.; Epple, M.; Surmenev, R.A. The structure of an RF-magnetron sputter-deposited silicate-containing hydroxyapatite-based coating investigated by high-resolution techniques. *Surf. Coat. Tech.* **2013**, *218*, 39–46. [CrossRef]
43. Surmeneva, M.A.; Surmenev, R.A.; Pichugin, V.F.; Chernousova, S.S.; Epple, M. In-vitro investigation of magnetron-sputtered coatings based on silicon-substituted hydroxyapatite. *J. Surf. Investig. X-Ray Synchrotron Neutron Tech.* **2011**, *5*, 1202–1207. [CrossRef]
44. Pichugin, V.F.; Surmeneva, M.A.; Surmenev, R.A.; Khlusov, I.A.; Epple, M. Study of physicochemical and biological properties of calcium phosphate coatings prepared by RF magnetron sputtering of silicon-substituted hydroxyapatite. *J. Surf. Investig. X-Ray Synchrotron Neutron Tech.* **2011**, *5*, 863–869. [CrossRef]
45. Surmeneva, M.; Tyurin, A.; Mukhametkaliyev, T.; Teresov, A.; Koval, A.; Pirozhkova, T.; Shuvarin, I.; Chudinova, E.; Surmenev, R. The effect of Si content on structure and mechanical features of silicon-containing calcium-phosphate-based films deposited by RF-magnetron sputtering on titanium substrate treated by pulsed electron beam. In Proceedings of the IOP Conference Series: Materials Science and Engineering, Tambov, Russia, 21–22 May 2015; p. 012028.
46. Wang, J.; Xue, C.; Zhu, P. Hydrothermal synthesis and structure characterization of flower-like self assembly of silicon-doped hydroxyapatite. *Mater. Lett.* **2017**, *196*, 400–402. [CrossRef]
47. Bohdansky, J. A Universal Relation for the Sputtering Yield of Monatomic Solids at Normal Ion Incidence. *Nucl. Instrum. Methods Phys. Res. Sect. B-Beam Interact. Mater. At.* **1984**, *2*, 587–591. [CrossRef]
48. Balas, F.; Perez-Pariente, J.; Vallet-Regi, M. In Vitro bioactivity of silicon-substituted hydroxyapatites. *J. Biomed. Mater. Res.* **2003**, *66*, 364–375. [CrossRef]
49. Okada, K.; Kameshima, Y.; Yasumori, A. Chemical shifts of silicon x-ray photoelectron spectra by polymerization structures of silicates. *J. Am. Ceram. Soc.* **1998**, *81*, 1970–1972. [CrossRef]
50. Steveson, M.; Arora, P.A.; Smart, R.S.C. XPS studies of low temperature plasma-produced graded oxide-silicate-silica layers on titanium. *Surf. Interface Anal.* **1998**, *26*, 1027–1034. [CrossRef]
51. Botelho, C.M.; Lopes, M.A.; Gibson, I.R.; Best, S.M.; Santos, J.D. Structural analysis of Si-substituted hydrooxyapatite:zeta potential and X-ray photoeletron spectroscopy. *J. Mater. Sci. Mater. Med.* **2002**, *13*, 1123–1127. [CrossRef]
52. Arcos, D.; Sanchez-Salcedo, S.; Izquierdo-Barba, I.; Ruiz, L.; Gonzalez-Calbet, J.; Vallet-Regi, M. Crystallochemistry, textural properties, and in vitro biocompatability of diffferent silicon-doped calcium phoshates. *J. Biomed. Mater. Res.* **2006**, *78*, 762–771. [CrossRef]
53. Long, J.D.; Xu, S.; Cai, J.W.; Jiang, N.; Lu, J.H.; Ostrikov, K.N.; Diong, C.H. Structure, bonding state and in-vitro study of Ca-P-Ti film deposited on Ti6Al4V by RF magnetron sputtering. *Mater. Sci. Eng. C* **2002**, *20*, 175. [CrossRef]

54. Van Raemdonck, W.; Ducheyne, P.; De Meester, P. *Metal and Ceramic Biomaterials*; CRC Press: Boca Raton, FL, USA, 1984; pp. 143–146.
55. Lo, W.J.; Grant, D.M. Hydroxyapatite thin films deposited onto uncoated and (Ti,Al,V)N-coated Ti alloys. *J. Biomed. Mater. Res.* **1999**, *46*, 408–417. [CrossRef]
56. Boyd, A.R.; Burke, G.A.; Duffy, H.; Cairns, M.L.; O'Hare, P.; Meenan, B.J. Characterisation of calcium phosphate/titanium dioxide hybrid coatings. *J. Mater. Sci. Mater. Med.* **2007**, *19*, 485–498. [CrossRef] [PubMed]
57. Astala, R.; Calderin, L.; Yin, X.; Stott, M.J. Ab initio simulation of Si-doped hydroxyapatite. *Chem. Mater.* **2006**, *18*, 413–422. [CrossRef]
58. Pan, Y.; Fleet, M.E. Substitution Mechanisms and Controlling Factors. In *Reviews in Mineralogy and Geochemistry -Phosphates -Geochemical, Geobiological and Materials Importance*; Kohn, M.J., Hughes, J.M., Eds.; American Chemical Society: Washington, DC, USA, 2002; p. 742.
59. Vandiver, J.; Dean, D.D.; Patel, N.; Botelho, C.M.; Best, S.M.; Santos, J.D.; Lopes, M.A.; Bonfield, W.; Ortiz, C. Silicon addition to hydroxyapatite increases nanoscale electrostatic, van der Waals and adhesive interactions. *J. Biomed. Mater. Res.* **2006**, *78*, 352–363. [CrossRef]
60. Takeda, S.; Fukawa, M. Role of surface OH groups in the surface chemical properties of metal oxide films. *Mater. Sci. Eng. B* **2005**, *119*, 265–267. [CrossRef]
61. Chen, W.; Liu, Y.; Courtney, H.S.; Bettenga, M.; Argrawal, C.M.; Bumgardner, J.D.; Ong, J.L. In vitro anti-bacterial and biological properties of magnetron co-sputtered silver-containing hydroxyapatite coating. *Biomaterials* **2006**, *27*, 5512–5517. [CrossRef]
62. Pietak, A.M.; Reid, J.W.; Stott, M.J.; Sayer, M. Silicon substitution in the calcium phosphate bioceramics. *Biomaterials* **2007**, *28*, 4023–4032. [CrossRef]
63. Gao, J.; Wang, M.; Shi, C.; Wang, L.; Wang, D.; Zhu, Y. Synthesis of trace element Si and Sr codoping hydroxyapatite with non-cytotoxicity and enhanced cell proliferation and differentiation. *Biol. Trace Elem. Res.* **2016**, *174*, 208–217. [CrossRef]
64. Reffitt, D.M.; Ogston, N.; Jugdaohsingh, R.; Cheung, H.F.J.; Evans, B.A.J.; Thompson, R.P.H.; Powell, J.J.; Hampson, G.N. Orthosilicic acid stimulates collagen type 1 synthesis and osteoblastic differentiation in human osteoblast like cells in vitro. *Bone* **2003**, *32*, 127–135. [CrossRef]
65. Xynos, I.D.; Edgar, A.J.; Buttery, L.D.K.; Hench, L.; Polak, J.M. Gene-expression profiling of human osteoblasts following treatment with the ionic products of Bioglass 45S5 dissolution. *J. Mater. Res.* **2001**, *55*, 151–157. [CrossRef]
66. Burling, L. *Novel Phosphate Glasses for Bone Regeneration Applications*; The University of Nottingham: Nottingham, UK, 2005.
67. Knabe, C.; Klar, F.; Fitzner, R.; Radlanski, R.J.; Gross, U. In vitro investigation of titanium and hydroxyapatite dental implant surfaces using a rat bone marrow stromal cell culture system. *Biomaterials* **2002**, *23*, 3235–3245. [CrossRef]
68. Stupack, D.G.; Cheresh, D.A. Get a ligand, get a life: Integrins, signalling and cell survival. *J. Cell Sci.* **2002**, *115*, 3729–3738. [CrossRef] [PubMed]
69. Miranti, C.K.; Brugge, J.S. Sensing the environment: A historical perspective on integrin signal transduction. *Nat. Cell Biol.* **2002**, *4*, 83–90. [CrossRef] [PubMed]
70. Anselme, K.; Bigerelle, M.; Noel, B.; Iost, A.; Hardouin, P. Qualitative and quantitative study of human osteoblast adhesion on materials with various surface roughnesses. *J. Biomed. Mater. Res.* **2000**, *49*, 155–166. [CrossRef]
71. Porter, A.E.; Patel, N.; Skepper, J.N.; Best, S.M.; Bonfield, W. Comparison of in vivo dissolution processes in hydroxyapatite and silicon-substituted hydroxyapatite bioceramics. *Biomaterials* **2003**, *24*, 4609. [CrossRef]
72. Thian, E.S.; Huang, J.; Best, S.M.; Barber, Z.H.; Bonfield, W. Magnetron co-sputtered silicon-containing hydroxyapatite thin films–an in vitro study. *Biomaterials* **2005**, *26*, 2947–2956. [CrossRef]
73. Thian, E.S.; Huang, J.; Best, S.M.; Bonfield, W. Fabrication of hydroxyapatite ceramics via a modified slip casting route. *Key Eng. Mater.* **2004**, *254*, 111–114. [CrossRef]
74. Hing, K.A.; Revell, P.A.; Smith, N.; Buckland, T. Effect of silicon level on rate, quality and progression of bone healing within silicate-substituted porous hydroxyapatite scaffolds. *Biomaterials* **2006**, *27*, 5014. [CrossRef]

75. Porter, A.E.; Botelho, C.M.; Lopes, M.A.; Santos, J.D.; Best, S.M.; Bonfield, W. Ultrastructural comparison of dissolution and apatite precipitation on hydroxyapatite and silicon-substituted hydroxyaptite in vitro and in vivo. *J. Biomed. Mater. Res.* **2004**, *69*, 670–679. [CrossRef]
76. Lo, W.J.; Grant, D.M.; Ball, M.D.; Welsh, B.S.; Howdle, S.M.; Antonov, E.N.; Bagratashvili, V.N.; Popov, W.K. Physical, chemical, and biological characterization of pulsed laser deposited and plasma sputtered hydroxyapatite thin films on titanium alloy. *J. Biomed. Mater. Res.* **2000**, *50*, 536–545. [CrossRef]
77. *Biomaterials Science: An Introduction to Materials in Medicine*; Ratner, B.D.; Hoffman, A.S.; Schoen, F.J.; Lemons, J.E. (Eds.) Academic Press: San Diego, CA, USA, 1996.
78. Scotchford, C.A.; Ball, M.; Winkelmann, M.; Voros, J.; Csucs, C.; Brunette, D.M.; Danuser, G.; Textor, M. Chemically patterned, metal-oxide-based surfaces produced by photolithographic techniques for studying protein- and cell-interactions. II: Protein adsorption and early cell interactions. *Biomaterials* **2003**, *24*, 1147. [CrossRef]
79. Dalby, M.J.; Pasqui, D.; Affrossman, S. Cell response to nano-islands produced by polymer demixing: A breif review. *IEE Proc. Nanobiotechnol.* **2004**, *151*, 53–61. [CrossRef] [PubMed]
80. Curtis, A.; Wilkinson, C. Nanotechniques and approaches in biotechnology. *Trends Biotechnol.* **2001**, *19*, 97–101. [CrossRef]
81. Kahn, S.P.; Auner, G.G.; Newaz, G.M. Influence of nanoscale surface roughness on neural cell attachement on silicon. *Nanomed. Nanotechnol. Biol. Med.* **2005**, *1*, 125–129. [CrossRef]
82. Ward, B.C.; Webster, T.J. The effect of nanotopography on calcium and phosphorus deposition on metallic materials in vitro. *Biomaterials* **2006**, *27*, 3064–3074. [CrossRef]
83. Dalby, M.J.; Giannaras, D.; Riehle, M.O.; Gadegaard, N.; Affrossman, S.; Curtis, A.S.G. Rapid fibroblast adhesion to 27 nm high polymer demixing nano-topography. *Biomaterials* **2004**, *25*, 77–83. [CrossRef]
84. Dalby, M.J.; Childs, S.; Riehle, M.O.; Johnstone, H.J.H.; Affrossman, S.; Curtis, A.S.G. Fibroblast reaction to island topography:changes in cytoskeleton and morphology with time. *Biomaterials* **2003**, *24*, 927–935. [CrossRef]
85. Ong, J.L.; Prince, C.W.; Lucas, L.C. Cellular reposonse to well-characterized calcium phosphate coatings and titanium surfaces in vitro. *J. Biomed. Mater. Res.* **1995**, *29*, 165–172. [CrossRef]
86. Wang, C.; Duan, Y.; Markovic, B.; Barbara, J.; Howlett, C.R.; Zhang, X.; Zreiquat, H. Phenotypic expression of bone-related genes in osteoblasts grown on calcium phosphate ceramics with different pase compositions. *Biomaterials* **2004**, *25*, 2507–2514. [CrossRef]

Review

Biomimetic Coatings Obtained by Combinatorial Laser Technologies

Emanuel Axente [1], Livia Elena Sima [2] and Felix Sima [1,*]

[1] Center for Advanced Laser Technologies (CETAL), National Institute for Laser, Plasma and Radiation Physics (INFLPR), 409 Atomistilor, Magurele 077125, Romania; emanuel.axente@inflpr.ro

[2] Department of Molecular Cell Biology, Institute of Biochemistry of the Romanian Academy, 296 Splaiul Independentei, Bucharest 060031, Romania; lsima@biochim.ro

* Correspondence: felix.sima@inflpr.ro; Tel.: +4-021-457-4243

Received: 13 April 2020; Accepted: 7 May 2020; Published: 9 May 2020

Abstract: The modification of implant devices with biocompatible coatings has become necessary as a consequence of premature loosening of prosthesis. This is caused mainly by chronic inflammation or allergies that are triggered by implant wear, production of abrasion particles, and/or release of metallic ions from the implantable device surface. Specific to the implant tissue destination, it could require coatings with specific features in order to provide optimal osseointegration. Pulsed laser deposition (PLD) became a well-known physical vapor deposition technology that has been successfully applied to a large variety of biocompatible inorganic coatings for biomedical prosthetic applications. Matrix assisted pulsed laser evaporation (MAPLE) is a PLD-derived technology used for depositions of thin organic material coatings. In an attempt to surpass solvent related difficulties, when different solvents are used for blending various organic materials, combinatorial MAPLE was proposed to grow thin hybrid coatings, assembled in a gradient of composition. We review herein the evolution of the laser technological process and capabilities of growing thin bio-coatings with emphasis on blended or multilayered biomimetic combinations. These can be used either as implant surfaces with enhanced bioactivity for accelerating orthopedic integration and tissue regeneration or combinatorial bio-platforms for cancer research.

Keywords: bio-coatings; biomimetics; laser deposition; PLD; MAPLE; tissue engineering; cancer

---

1. Biomimetic Materials

During the last two decades, new challenges in nanoscience and nanotechnology have been continuously addressed, in particular within the biomedical field [1–3]. The necessity of new biomaterials with improved properties have led to interdisciplinary studies at the interface between physics, chemistry, materials science, biology, and medicine [1]. There are, by now, four generations of biomaterials with the latter one capable of adapting to extra- and intra-cellular processes that could allow for the understanding of signaling pathways mediating inter- and intra-cellular communications [4–6]. However, there is still a breach in the in-depth understanding of nanomaterial interactions with biological entities both in vitro and in vivo. The design of innovative biomaterials should aim to precisely control the composition–properties relationship in order to modulate cell behavior in the field of tissue engineering and regenerative nanomedicine [7] or cancer theranostics [8].

With the development of the materials science field, biomaterial surface properties have advanced from bio-inert to bio-active and bio-resorbable, then to bio-conductive. These characteristics have been shown to strongly influence cell behaviors such as viability, proliferation, migration, and differentiation [4,5,9]. Biomimetic materials are, in addition, considered to effectively control cell behavior through partial replication of specific tissue features [10]. In a simple description, biomimetics rely on the ability to create synthetic characteristics from nature-inspired shapes and principles with the aim to achieve the desired biological responses. As a consequence, an improvement in the biomaterial surface could be attained by modifying their basic morphological properties to create a biomimetic environment that could eventually control the cell–surface interaction. In addition, in order to structurally and functionally recapitulate the natural micro-environment, inorganic–organic composite structures become necessary. Hence, a broad variety of coating materials are extensively explored for tissue engineering applications, spanning from ceramics, natural and synthetic biopolymers, proteins, peptides, enzymes, and growth factors to cover *composite biomimetic materials* consisting of blends with drugs or other biomolecules.

2. Biomimetic Coatings: In Vitro Testing Strategies and Clinical Trials

This section focuses on tissue engineering applications of bio-coatings, although few examples will highlight cancer research. The main materials used in implantology are metals, ceramics, composites, and polymers. In the case of prosthetic and dental applications, the uses of bio-coatings are anticipated to improve osseointegration of metal implants, while preventing allergies and chronic inflammation. The main culprit for these clinical manifestations is the corrosion of metals and alloys used in clinics for orthopedic, orthodontic, cardiovascular, or craniofacial metallic implant fabrication (mainly titanium-based alloys). The main techniques used to enhance the corrosion resistance of biomaterials were reviewed by Asri et al. [11].

First attempts at covering the surface of stainless-steel implants with hydroxyapatite (HA, $Ca_{10}(PO_4)_6(OH)_2$) by electrochemical deposition were contradictory and somehow disappointing in terms of resistance to corrosion [12,13]. This bioactive material has, however, the advantage of resemblance to the inorganic chemical structure of bone and teeth. Therefore, numerous studies have been dedicated to enhance biomaterial properties by coating inert metallic substrates with HA, ion-substituted calcium phosphates (CaPs), or hybrid organic-inorganic coatings containing CaPs [14].

2.1. In Vitro Testing of Inorganic Coatings for Tissue Engineering

CaPs are not osteoinductive (able to induce *de novo* heterotopic bone formation) *per se* unless they are structured in porous structures [15]. It has been suggested that such topography induces osteogenesis by regulating primary cellular cilia length and transforming growth factor (TGF) receptor recruitment [16]. Recently, molecular cell analyses have revealed that plasma cell glycoprotein 1 (PC-1) is a key protein responsible for the osteoinductive response of cells to CaP ceramics as a negative regulator of bone morphogenic protein-2 (BMP-2) signaling, a key pathway governing bone development and regeneration [17]. The proposed molecular driver of osteoinductivity is the localized depletion of calcium and phosphate ions from body fluid [18].

It is now accepted that composition, stiffness, and topography are key biophysical features that modulate bone progenitor cell differentiation [19]. Different processing techniques have been tested for their capacity to induce the proper bone formation response in osteoblasts and mesenchymal stem cells (MSCs) when exposed to CaP-based ceramic coatings and composites thereof.

A comparison between plasma electrolytically oxidized (PEO) and plasma-sprayed HA coatings on a Ti-6Al-4V alloy revealed that the PEO obtained films induced a significant increase in collagen production by human MG-63 osteoblast-like cells, while they had 78% lower surface roughness compared to plasma spraying [20]. Recent research has helped in successfully overcoming the low adhesive bond strength of the plasma sprayed HA coating over metallic substrates by creating an interfacial layer consisting of a gradient HA coating prepared by laser engineered net shaping (LENS™) followed by plasma spray deposition [21].

Researchers have next tested the possibility of substituting the calcium ions in HA or other CaP by biologically relevant ions or compounds. These are known to provide support for both osteogenesis and angiogenesis processes [22]. Studies led by Adriana Bigi et al. proposed and tested in vitro several doping solutions for CaP-based laser deposited coatings containing either strontium (Sr) alone [23] or in combination with magnesium (Mg) [24] or with the drug, zoledronate (ZOL) [25]. For this work, different approaches were employed for coating synthesis and further subjected to in vitro testing with osteoblasts and osteoclasts. These studies have shown an improvement in osteoblast adhesion, proliferation, and differentiation, while osteoclast proliferation decreased with an increase in Sr content of the Sr-HA PLD coatings [23]. Mg and Sr doping of octacalcium phosphate (OCP) enhanced osteoblast proliferation, activity, and differentiation [24]. Testing of reciprocal gradients of Sr and ZOL doped HA in osteoblast-osteoclast co-cultures has revealed associative conditions that favor bone production activity more than each of the dopants alone [25]. Experiments showed that ZOL promotes collagen type I (COLL1) production, whereas Sr significantly increases the production of alkaline phosphatase (ALP), two osteoblast markers expressed during late and early differentiation, respectively. Sr was more effective in enhancing osteoblast viability and activity, while ZOL was more effective in inhibiting osteoclastogenesis. The laser method allowed, in this case, the modulation of the coatings' impact on bone cell activity by "titrating" the gradient proportion in which the two molecules produce most beneficial outcome, with minimal cytotoxicity.

Inductively coupled radio frequency (RF) plasma spray was also used to generate Sr–HA and Mg–HA coatings on Ti. Their performance was investigated using human fetal osteoblasts and results showed an increased cell attachment and proliferation as well as improved differentiation in the presence of Sr when compared to HA and Mg-HA [26].

In vivo experiments in a rat model of osteoporosis have confirmed the improved implant osseointegration when Ti was coated by Sr-doped HA when compared to Zn and Mg dopants [27]. Sr–HA bonding with animal bone was almost twice as strong as for HA alone, as proven by the biomechanical tests. This result was supported by the increased new bone formation surrounding the implant, when HA contained ion dopants. The coating was obtained, however, through electrochemical deposition; there are no similar in vivo reports on laser obtained Sr–HA coatings. Similar improvements in bone regeneration were reported for 5.7% Mg–HA when tested on New Zealand White rabbits using the granulate as filling for a femoral bone defect compared to HA alone [28].

Corrosion is not the only problem related to implant success. Another pathophysiologic event that prevents long-term stability of the implanted biomaterial is the loosening of the prosthesis due to its encapsulation by a fibrotic scarring tissue. Therefore, besides the coatings used to prevent metallic corrosion, solutions have been developed to increase the osseointegration of implants, especially those addressing long bones replacement, where forces applied on the connecting joints are higher and increase bonding strength to compensate them is much needed. Mechanical stimulation is impeded at the implantation site, which negatively impacts new bone formation by the surrounding tissue. This blockage in mechanotrasduction, due to decreased mechanical loading, is known as "stress shielding" and is reduced by the use of flexible materials in the fabrication of implant stems [29].

A representative study for this approach is the one proposed by Scarisoreanu et al. [30]. In this work, the laser synthesis of a functional biocompatible piezoelectric material coating on a flexible Kapton polymer substrate is reported. The experiments validate such structures as optimal support for MSC adhesion, proliferation, and osteogenic differentiation, which is in favor of potential future use of piezoceramic coatings on bone implants.

Efforts to predict the fibrotic tendency of a biomaterial through a comparative multi-parametric in vitro model have revealed a high correlation between an in vitro test based on autologous plasma to in vivo-obtained biomaterial assessments [31]. It is foreseen that a combination of a three-dimensional fibrin matrix and primary macrophage assays would help identify promising biomaterials and decrease the need for animal studies.

2.2. *In Vitro Testing of Organic Coatings for Tissue Engineering*

Osteoprogenitor cells response to their microenvironment is largely mediated by signals received from the extracellular matrix (ECM) (reviewed in [32]) that are transmitted to the cell nucleus *via* actionable signaling pathways. Hence, stem cell fate is modulated epigenetically upon integration of signals from outside the cell. Recent findings indicated that this mechanism can be coerced into diverting stem cell commitment to desired differentiation pathways: biochemical pre-treatment of MSCs with either of the five identified epigenetic modulators (Gemcitabine, Decitabine, I-CBP112, Chidamide, and SIRT1/2 inhibitor IV) enhanced osteogenesis in vitro [33]. Out of these, Gemcitabine and Chidamide successfully rescued the loss in osteogenic differentiation potential in MSCs obtained from aged donors, hence providing indication of the potential use of these small molecules as pharmacological agents against bone demineralization.

Similar effects are searched for when screening different biomaterials for prosthetics/bone regeneration purposes. In this case, the interaction of MSCs with the biomaterial surface is mediated by mechanotransduction pathways that engage cell integrin receptors binding to ECM molecules adsorbed on the substrate. The ideal implant biomaterial would provide the necessary cues in favor of osteogenic differentiation while discouraging cell commitment alternatives.

The *top–down* approach of generating scaffolds from tissues by decellularization and/or intermediary ECM purification steps largely preserves the complex composition of the in vivo niche, while the *bottom–up* synthetic approach uses key elements of the bone tissue able to induce osteogenic commitment in order to support the process of endochondral ossification involved in bone healing [34,35]. While the first strategy is fitted mainly for filling bone defects, the second is amenable to the development of biomimetic implant coatings. A purely biomaterial-based solution for the induction of endochondral ossification proposed for the challenging regeneration of critical-size defects was recently reported for the first time [36]. This involves the use of porcine collagen as a scaffold with channel-like pore architecture to assist the first steps of bone healing through extracellular matrix alignment, $CD146^{+}$ progenitor cell accumulation, and restrained vascularization. The recapitulation of developmental bone growth process was demonstrated in human stromal cell culture and in rat models of bone regeneration, where the macroporous scaffold functioned as a highly aligned biomaterial template that aligns ECM fibers along the bone axis. Alternatively, a synthetic ECM was also proposed as a hybrid between hyaluronan–fibrin and a polymer biodegradable template represented by Poly(Lactic-co-Glycolic Acid) (PLGA) microspheres [37].

In order to create coatings with increased bone tissue biomimetism, synthetic polymers or proteins are tested in conjugation or not with inorganic compounds. They are either biodegradable [38], and designed to provide temporary support for the regenerating tissue, or stable through the whole healing process [39]. Bioactive composite scaffolds combining nanofibrous polycaprolactone (PCL) matrix and HA or TCP as the mineral phase, either added inside the fiber structure or as a film obtained by electrospinning, represent a biomimetic material able to induce MSC osteogenesis, without the need of osteoinductive factor treatment [40]. Furthermore, carbonated HA–gelatin composite-coated PCL/TCP scaffolds were shown to stimulate osteogenesis compared to PCL/TCP alone [41]. ECM-mimetic scaffolds [42] and coatings [43] have been designed that contain growth factors or ECM proteins to be released in the healing environment upon biomaterial implantation in order to enhance tissue mineralization.

2.3. *In Vitro Testing of Bio-Coatings for Cancer Research*

Aside from the extensive research and development in bone tissue engineering field to implement the use of biomaterial coatings for specific needs, there are also a few recent examples of use in cancer research.

The development of biomimetic constructs by laser direct-write has allowed the proposal of a model to study breast cancer cell invasion into adipose tissue [44]. The construct reflects mammary microenvironment architecture through replication of the spatial relationship between cancer cells and tissue resident adipocytes, which are encapsulated in alginate–collagen microbeads. This tissue-like structure enables the investigation of the physiological contribution of obesity to breast cancer cell invasion. Another example represents a potential cell-based cancer immunotherapy application of biomaterials. For this application, magnetron sputtering was used to produce micropatterned nickel titanium thin films loaded with ovarian cancer-specific CAR-T cells for local delivery into solid tumors to eradicate established multifocal disease [45].

It is rather accepted that none of the deposition techniques depicted above could fabricate an "*ideal coating*", able simultaneously to: (i) precisely mimic the native physiological micro-environment, and (ii) accurately control drugs or other functional molecule delivery from coatings to specific sites. Accordingly, the biomaterials research community has focused on the development and the fabrication of blended and multilayered bio-coatings in view of obtaining multiple functionalities, eventually required in clinical trials [46–49].

2.4. In Vivo Clinical Trials for Regenerative Medicine

Despite the multitude of research approaches to enhance orthopedic and dental implant osseointegration, the majority of those that reached the clinical trials stage are based on CaP coatings (recently reviewed in [50]). The main physical deposition techniques to produce CaP-based coatings are the thermal spray processes followed by other vaporization-based methods (PLD, MAPLE, ion-beam-assisted deposition (IBAD), RF magnetron sputtering) [51].

A search using the "bone implant coatings" keywords in the *ClinicalTrails.gov* database retrieved 71 entries. Out of these, 29 were completed clinical trials, while the others are still on-going. Three trials have been completed with results, out of which two tested plasma sprayed coated biomaterials. First is an HA plasma sprayed acetabular cup that was tested in parallel with a BoneMaster HA coated cup to compare the effect upon bone density and clinical outcomes in patients with total hip replacement (identifier NCT00859976). After two years post-operation, two patients in the plasma-sprayed shell receiving group had not fixated implants and another one showed an unstable implant out of 22 participants, while no radiolucency was revealed in the BoneMaster treated patients ($n = 12$). The other clinical assessments showed no significant differences between patient outcomes in the two groups. Second was a new Bone Anchored Hearing Aid (BAHA) device that was tested, consisting of a commercially pure titanium coated with a plasma sprayed HA layer on the entire soft tissue-contacting surface of the abutment up to 3 mm below the top surface. The randomized controlled trial took place in four European countries and was aimed at comparing this new bone conducting BA400 model implant with the traditional BA300 abutment (identifier NCT01796236, [52]). The use of the HA coated abutment allowed for soft tissue preservation and improved cosmetic results for patients as well as demonstrated cost savings. Third was a dental bone inductive implant obtained by adsorption of rhBMP2 on a porous titanium oxide surface (identifier NCT00422279, [53]). Four participants presenting alveolar ridge abnormality were enrolled in the study bearing two implants each. The aim was to evaluate implant stability and new bone formation around the implant. Based on preliminary experiments in canine models, a minimum dose of 15 and 30 μg rhBMP2 per implant was chosen. The assessment of implant stability at six months showed positive results for 4/4 implants in supra-alveolar position while 3/4 implants were stable upon implantation at extraction sites. However, no evidence of bone growth was seen in any of the cases, which rendered the study not successful.

Continuous efforts are still necessary for testing the various promising prosthetic materials in a clinical setting for the validation of their patient benefits.

3. Biomimetic Coatings: Advantages and Drawbacks of Processing Technologies

3.1. Biomimetic Processing Technologies

There are various physical and chemical deposition techniques employed for the synthesis of bio-coatings and they exhibit advantages and limitations regarding the type of material, the preservation of stoichiometry, control of morphological and structural properties, or processing of the coating area. Among them, plasma vapor deposition (PVD) techniques, usually apply to inorganic material coating. One may enumerate thermal evaporation, atomic layer deposition, electron beam evaporation, sputter deposition in plasma, reactive sputter deposition, cathodic arc deposition, pulsed electron deposition, electroless plating, or laser deposition. On the other hand, there are also coating methods appropriate for organic materials such as the Langmuir–Blodgett dip coating, sol-gel, layer by layer deposition, aerosol spraying, dip coating, spin coating or laser evaporation that entail liquid solutions of the material in a volatile solvent. Typically, each method is suitable to a limited class of compounds, either inorganic or organic. This represents a limiting factor, particularly in the case of a composite bio-coating. Composites become necessary to solve specific medical problems such as to improve osseointegration or to decrease local tissue inflammation and avoid infection.

We reviewed several studies (Table 1) devoted to the deposition techniques for either inorganic, organic, or composite materials.

Table 1. The main deposition methods of inorganic, organic, and hybrid biocompatible coatings selected from the last years.

Deposition Method	Advantages	Drawbacks	Bio-Coating Materials/Application
		Laser Techniques	
Pulsed laser deposition (PLD)	Stoichiometric and adherent coatings, morphology control, easy to obtain multilayered thin films, good versatility of experimental design, thickness control.	Limited to inorganic coatings, small area covering (few cm), high costs. Micrometer-sized droplets and particulates on surfaces.	Inorganic coatings: HA [54,55], and Bioglass (BG) [56]/Implant devices.
Combinatorial-Pulsed laser deposition (C-PLD)	Preserve the properties of PLD, synthesis of combinatorial libraries of thin films, controlled doping of coatings, cover larger substrates (glass slide).	Limited to inorganic coatings, high costs.	Inorganic coatings: Ag-doped HA [57], Ag-doped Carbon [58]/Model surfaces.
Matrix assisted pulsed laser evaporation (MAPLE)	Applied to both organic and inorganic coatings, multilayers and multistructures, nanoparticulate films, thickness control.	Generation of micrometer-sized droplets and particulates on surfaces, small covering areas.	Inorganic, organic, hybrid coatings: HA [55], HA/Lactoferrin/polyethylene glycol-polycaprolactone copolymer [59], hybrid BG-biopolymer [60]/Implant devices, drug delivery.
Combinatorial-Matrix assisted pulsed laser evaporation (C-MAPLE)	Preserve the properties of MAPLE, single-step synthesis of combinatorial bio-coatings, suitable for organic, inorganic and composites, cover larger substrates (glass slide).	Generation of micrometer-sized droplets and particulates on surfaces, high costs.	Inorganic, organic, hybrid coatings: Chitosan/ bio-mimetic apatite [61], CaPs, biopolymers [62,63] Levan and Chitosan blends [64], Graphene Oxide, BSA protein, drugs [65]/Implant devices, drug delivery, model surfaces.
Laser-induced forward transfer (LIFT)	Micro-patterns with high spatial resolution, size and separation distance between structures, easy control.	Limited to patterns, difficulties for large area micro-fabrication, difficult to control the height of the patterns.	Inorganic, organic, hybrid micropatterns: collagen and nanoHA [66], proteins and biomaterials [67], mesenchymal stromal cells [68]/Implant devices, drug delivery, model surfaces.
		Non-Laser Techniques	
Radio frequency magnetron sputtering (RF-MS)	Stoichiometric transfer, uniform dense coatings, good versatility of experimental design.	Limited to inorganic coatings, amorphous coatings, rather expensive.	Inorganic coatings: CaP [69,70], BG [71–73]/Implant devices.
Pulsed electron deposition (PED)	Stoichiometric and adherent thin films, low cost compared to other PVD techniques.	Limited to inorganic coatings, small area covering.	Inorganic coatings: HA, CaP, biogenic CaP, BG [74]/Implant devices.
Plasma spray (PS)	Simplest, operates at low costs.	Use for inorganic coatings only, poor coating adhesion, weak bonding strength at ceramic-metal interface.	Inorganic coatings: HA [75,76], and BG [77]/Implant devices.

Table 1. *Cont.*

Deposition Method	Advantages	Drawbacks	Bio-Coating Materials/Application
Layer-by-layer (LBL)	Good thickness control, viscoelasticity/bioactivity, coat multiple substrates of all scales.	Poor layers adhesion to substrates, difficult to create multilayers due to solvent issues, difficult to generate gradient coatings.	Organic coatings: biomaterials for drug delivery [78]/Drug delivery, model surfaces.
Langmuir-Blodgett dip coating	Possibility to assemble monolayers, high spatial coverage.	Limited to very thin films, typically used for organic coatings only.	Organic coatings: PCL-Blend-PEG [79]/Drug delivery, model surfaces.
Adsorption on surface	Simple, rapid and inexpensive, useful for organic bio-coatings.	Poor adhesion on substrates, poor uniformity, difficult to create multi-layer assembling.	Organic coatings: BMP-2, BMP-6, BMP-7 with fibronectin onto HA coatings [80]/Implant devices, drug delivery.
Spin coating	Simple and inexpensive, uniform coatings, accurate thickness control, high spatial coverage.	Use for 2D surfaces only, typically used for organic coatings, solvent issue in case of multilayers, poor adherence.	Organic coatings: protein-polysaccharide thin films [81]/Implant devices, drug delivery.
Sol-gel (SG)	Available for both organic and inorganic coatings, simple operation, high versatility.	Poor adhesion to substrates, difficult to create multilayers due to solvent issues, difficult to generate gradient coatings.	Inorganic, organic, hybrid coatings: HA, BG [82], Gentamicin/Chitosan/BG composite [83]/Implant devices, drug delivery.
Electrophoretic deposition (EPD)	Useful for both organic and inorganic coatings, simple processing setup.	Difficult to create multilayers due to solvent issues, poor coating adhesion.	Inorganic, organic, hybrid coatings: HA–iron oxide–chitosan composite [84,85]/Implant devices, drug delivery.

3.2. *Bio-Coating Adhesion Issues*

Aside from specific characteristics of each technological process either applied to inorganic, organic, or composite materials, there is a strong correlation between the structure–property relationship and adhesion strength of synthesized coatings. Indeed, in the case of metallic implants for dentistry or orthopedics, a poor adherence of the covering thin layers could stand for premature delamination and eventually failure of the implant. For example, plasma-sprayed HA coatings contain stresses that induce a large mismatch between thermal expansion coefficients of the coating and metal substrate [86]. Typically, the adhesion strength of the bio-coatings is carried out by scratch, nanoindentation tests, and pull-off tests. Modifications of the substrate surface or the uses of an intermediate buffer layer play a significant role in the coating growth process and adhesion, particularly for inorganic materials. Specifically, sand-blasting, followed by acid-etching of a titanium surface, showed the influence of nano/micro-structure on grain size, mechanical properties, and surface wettability of Ag-doped HA bio-coatings deposited by the radio frequency magnetron sputtering (RF-MS) process [87]. In this study, nanoindentation tests revealed significantly higher nanohardness and Young's modulus values with decreasing grain size. The introduction of an intermediate buffer layer (TiN, ZrO_2, or Al_2O_3) between the Ti alloy substrate and HA coating deposited by PLD allowed for thin films with improved mechanical characteristics to be obtained compared with structures synthesized without the transitional layer [88]. In this case, the friction coefficient was found to be reduced by 25% when an intermediate layer was used. In addition, a direct comparison of RF-MS and PLD demonstrated that HA coatings exhibited improved mechanical properties when grown by both methods on Ti alloy substrates previously coated with a TiN buffer layer [89]. Generally, for HA coatings deposited by PLD, good adhesion properties are described when the substrates are heated during coating processes and samples are further thermally treated post-deposition [90]. However, pull-off tests revealed, in some cases, a decrease of tensile strength value with increasing substrate temperature from 480 to 550 °C during the deposition process [91]. A comprehensive comparative investigation on the adhesion of HA coatings obtained by different PVD techniques on Ti alloy substrates was reported by Mohseni et al. [92]. On the other hand, biopolymer adhesion to metallic substrates was found to be good in most cases [93]. Aside from material chemistry, surface energy was proven to be a dominant factor to induce in vitro cell adhesion and proliferation, and found to play a rather more important role than surface roughness for cell colonization onto engineered tissue scaffolds [94].

4. Biomimetic Laser Processing

Laser-based technologies have been demonstrated to be a clean, fast, and cost-effective alternative to the existing procedures for organic, inorganic, and composite bio-coatings. They are in the forefront of several discoveries in nanoscience and nanotechnology. Laser-based deposition techniques have been widely employed for the fabrication of various thin biomimetic composite coatings [7,63,95–98]. The book edited by Schmidt and Belegratis [99] presents a detailed state-of-the-art of the laser technologies until seven years ago as well as the materials that are typically employed in cutting-edge biomimetic applications.

Two conventional laser approaches used nowadays for assembling biomimetic thin coatings are: (i) pulsed laser deposition (PLD) for the synthesis of inorganic coatings and (ii) matrix assisted pulsed laser evaporation (MAPLE) for the fabrication of organic, inorganic, and hybrid organic–inorganic coatings. Both techniques use pulsed laser beams to ablate solid targets in vacuum, deposit the ablated material cluster by cluster on facing substrates, and form a thin coating. By readapting setup configurations, in particular employing multiple pulsed laser beams and/or multi-targets, combinatorial-PLD and combinatorial-MAPLE were developed to fabricate thin blended coatings with a gradient of composition in a single-step process. The key benefits and drawbacks of these techniques have been previously addressed in a comparison with other non-laser-based methods [7,62,63,97] and are briefly summarized in the following. Both methods are highly versatile, with individual process parameters able to control and tailor morphological and structural coating properties. The use of laser beams for target irradiation in vacuum allows for material processing in a contamination-free environment, suitable for biomedical applications.

4.1. Pulsed Laser Deposition

In case of PLD, the main advantages refer to the possibility of preserving the material stoichiometry during transfer, but also to switching the composition chemistry and structure to generate new properties. During laser irradiation, a plasma plume is formed. This plume drives the atoms, ions, and nanoparticles from the ablated target to a facing collector where a thin uniform coating grows additively. The thickness is controlled by the number of applied laser pulses. The synthesis of new materials can be achieved by introducing different gases and controlling the pressure inside the chamber. This is a reactive process that was named reactive-PLD (R-PLD) due to the chemical reactions in the plasma phase. Pioneering work reported the synthesis of TiN thin films by laser ablation of a Ti target in an N_2 gas that filled the reaction chamber [100]. Highly adherent thin coatings can be obtained due to the high kinetic energies of ablated species when hitting a heated substrate. Multilayered coatings with predefined thickness and/or doped structures can be eventually fabricated by a two-step process in which distinct targets are ablated successively [101].

4.2. Matrix Assisted Pulsed Laser Evaporation

MAPLE process uses milder conditions (e.g., laser energies one order of magnitude lower than in case of PLD), just above the threshold of frozen target vaporization. It therefore allows for the fabrication of either inorganic, organic, or hybrid coatings with a high versatility [59,96,97,102–105]. MAPLE offers specific benefits with respect to PLD, demonstrated by the safe transfer and deposition of delicate compounds such as proteins [106,107] or biopolymers, without impeding on the stability of their functional characteristics, which rely on preserving their nanoscale structure [103].

The MAPLE technique was first introduced twenty years ago by McGill and Chrisey [108] at the U.S. Naval Research Laboratory in order to attain damage-free pulsed laser evaporation of organic materials and thin films in the late 1990s. It was designed as an alternative to PLD in order to avoid organic material decomposition, degradation, and/or denaturation induced by high laser powers during ablation. Since then, MAPLE has evolved to produce a broad spectrum of organic, inorganic, and hybrid coatings [109–112] for various biomedical applications as well as for energy [113–115], sensing [116–118], wearable electronics, and photonic devices [98,119]. Indeed, although initially designed for polymers, the method proved successful for a large variety of compounds such as proteins [106,107], enzymes [120,121], polysaccharides [103,122,123], calcium phosphates [24,55,124,125], nanoparticles [126,127], proteins and drug functionalized graphene oxide [65], or carbon nanostructures embedded in organic matrices [128]. Its maturity has already been achieved and existing commercial installations fulfil the anticipations of the scientific community [129].

MAPLE is an additive physical vapor deposition process that was extensively explored for the functionalization of solid substrates with various coatings. In its most used configuration, the MAPLE procedure is based on pulsed laser irradiation of a cryogenic target using a UV laser beam (Figure 2). Typically, the target is composed of solute molecules that are dissolved in an appropriate solvent, compatible with the employed laser evaporation wavelength. During MAPLE processing, the frozen solute biomolecules are then transferred and assembled onto a receiving substrate, typically placed parallel with respect to the target at a specific separation distance (few centimeters in most cases). In the case of MAPLE, the evaporated biomolecules are collected on solid substrates, pulse by pulse, generating a thin coating with thicknesses from a few to several hundreds of nanometers. The targets are prepared by dissolving the active material in an appropriate solvent (e.g., water, chloroform, DMSO), followed by immersion of the solution in liquid nitrogen (LN) for an optimized time (typically 5–15 min, dependent on the solvent–solute pair type). After target solidification, this is placed inside a customized or commercial stainless steel reaction chamber. At this stage, the deposition parameters are set: target-to-substrates separation distance (3–5 cm), substrate temperature (room temperature or gentle heating), the nature and the pressure inside the reaction chamber (low vacuum *vs.* inert or reactive gas atmosphere). The most employed laser sources are excimer lasers, typically used for PLD. Setups and experiments in this review used excimer lasers (e.g., KrF^*, $\lambda = 248$ nm, with pulse duration $\tau_{FWHM} = 25$ ns, operated at $\upsilon = 1$–40 Hz), but other laser sources have also proven successful for obtaining functional organic coatings (Nd:YAG pulses at $\lambda = 266$ nm [105,114,118], $\lambda = 355$ nm [117], or for resonant absorption by Er:YAG laser at $\lambda = 2.94$ μm [130,131]. A critical step before the deposition protocol, mainly for biomedical applications, involves careful substrate cleaning in successive ultrasonic baths of acetone, ethanol, and deionized water (at least 15 min each).

Aside from deposition parameter control (laser wavelength, pulse duration, laser energy, laser spot focusing, repetition rate, target-to-substrate separation distance, substrate temperature, ambient conditions, number of applied laser pulses) one may also design distinct substrate arrangements in various geometries. The literature suggests both on-axis and off-axis configurations [132], which are necessary to precisely control the thickness and spreading of the coating. Such geometries should be further correlated with substrate positioning with respect to target rotation (Figure 2).

Two setup configurations can be used: (i) samples with the same composition and thickness are obtained by rotating both the target and the substrates (Figure 2a), and (ii) samples with compositional gradient and distinct thicknesses are prepared by rotating the target and keeping the substrates fixed (Figure 2b). In the latter case, the main evaporation flux is perpendicular to the closest sample C1, while a concentration gradient toward C4 is naturally achieved due to material spreading along the radial–orthogonal direction of the substrates (x direction in Figure 2b). The goal of adopting such irradiation geometry is necessary when the generation of compositional gradient coatings is applied in a single-step process. One possible application revealed a possibility of investigating a drug dose-dependent effect when incorporated into the coatings, with respect to controlled delivery for cancer studies [65].

Indeed, it is generally accepted that when compared to other conventional, non-laser deposition methods (e.g., drop-casting, spin-coating, dip-coating, Langmuir–Blodgett), the MAPLE technique allows for high experimental versatility and somehow high control of coating thickness challenging also ultrathin film structures. Other important advantages refers to the possibility of preserving structural and functional properties, even for very delicate compounds. The method was shown to easily provide congruent transfer of adherent and uniform coatings on centimeter sized substrates, micro-fabrication of multilayers from a multi-target system, and the possibility of fabricating gradient thin films in a single-step process. However, the main drawbacks are still related to difficulties met when large-area coatings (tens of centimeters) should be uniformly covered.

UV-MAPLE alternatives such as Resonant Infrared- (RIR-) and emulsion-based RIR-MAPLE including fundamental phenomena of laser beam interaction with frozen target and various applications have been previously reviewed elsewhere [98]. Briefly, the goal of using infrared (IR) wavelengths is to adjust the target absorption to a resonant region specific to the molecules of the solvent matrix, thus minimizing laser interaction with the materials to be transferred. Indeed, most of the vibrational frequencies of organic solvents are in the infrared region, and thus IR lasers can represent a viable alternative to UV lasers. Here, the IR excitation could be easily adapted for resonant interaction with specific chemical vibrational modes within the most used solvent matrix, and consequently, organic molecules experience minimal photochemical and structural degradation. A simplified schematic representation of the laser interaction processes applied to polymers using PLD, UV-MAPLE, RIR-MAPLE, and emulsion-based RIR-MAPLE techniques is presented in Figure 1.

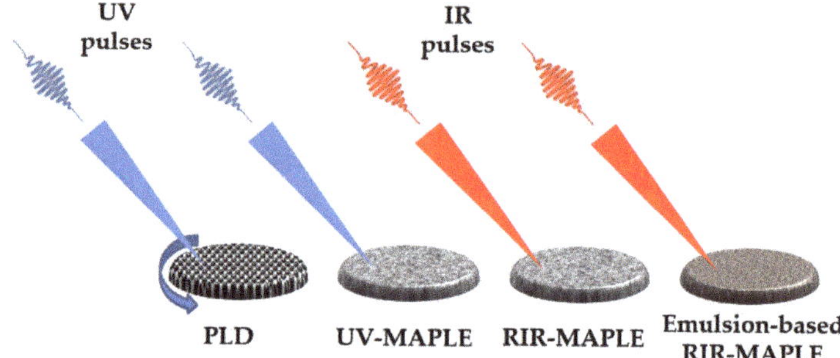

Figure 1. Simplified schematic sketch of the main laser-based deposition techniques of polymers: pulsed laser deposition (PLD), ultraviolet-matrix assisted pulsed laser evaporation (UV-MAPLE), resonant infrared-matrix assisted pulsed laser evaporation (RIR-MAPLE), and emulsion-based RIR-MAPLE techniques when compared to laser wavelengths (UV *vs.* IR) and target composition (solid pellet in case of PLD *vs.* cryogenic targets of frozen solutions/emulsions in case of MAPLE). Reprinted with permission from [98]. Copyright 2017 AIP.

The main differences between the processes are related to the physical–chemical processes at the laser–target interaction level. Polymer targets used with PLD typically consists of pressed powder or pellets, while MAPLE targets are composed of frozen solution. In the particular case of emulsion-based RIR-MAPLE, oil-in-water and/or water-in-oil emulsions are used in correlation with the hydrophobic/hydrophilic nature of the polymers to be grown as thin coatings. Few molecular dynamics simulation studies have addressed both UV- and RIR-MAPLE [133–135]. Another coarse-grained chemical reaction model [136] and a semi-empirical model based on the thermodynamics and kinetics of phase transitions in frozen solvent matrices [137] were evaluated for the complex transfer of organic molecules in MAPLE process.

5. Bio-Coatings with Multilayer Configurations and Gradient of Composition by Laser Deposition

In this section, we present a few representative examples of composite biomimetic coatings fabricated by laser techniques. Sima et al. [96] showed that multilayered inorganic–organic thin implant coatings could be obtained in a two-step laser process. PLD was first employed for growing HA thin films onto titanium substrates. Furthermore, fibronectin (FN) coatings were deposited by MAPLE on top of HA layers. Using a cryogenic temperature-programmed desorption mass spectrometry analysis, the authors found that less than 7 μg FN per cm^2 HA surface is an optimum concentration for improving adhesion, spreading, and differentiation of osteoprogenitor cells. Moreover, the possibility of using gradient coatings to study drug delivery influence when interacting with melanoma cells was evidenced [65]. BSA-functionalized graphene oxide nanomaterials (GONB) incorporating inhibitory drugs demonstrated an efficient dose-dependent effect on melanoma cells.

5.1. Combinatorial Laser Technologies

Combinatorial MAPLE (C-MAPLE) was first introduced in 2012 by Sima et al. [138] in order to fabricate coatings with compositional gradients on large substrate areas (several cm), in a single-step process. The method could be considered an *extension* of combinatorial-PLD (C-PLD), initially introduced by Takeuchi et al. [139], for the fabrication of inorganic compositional library thin films. Few C-PLD experiments aimed to generate inorganic bioactive coatings with gradient of composition for tissue engineering have been reported [57,58]. Socol et al. [57] have shown the possibility of controlling the composition and surface morphology characteristics in the case of Ag-doped CaP thin films by using C-PLD. Later, the same group reported the synthesis of antimicrobial libraries of Ag-doped carbon thin films, in order to be proposed as coatings that could minimize the risk of implant-associated infections [58].

C-MAPLE proved an appropriate method for the synthesis of several bio-coatings with compositional gradients and morphology for either organic, inorganic, or hybrid nanostructures with the view of creating biomimetic microenvironments. Representatives examples refer to protein embedded in biodegradable polymers [43], polysaccharides [64,123], ion and drug doped CaPs [25,61,140], or graphene oxide for controlled drug release [65].

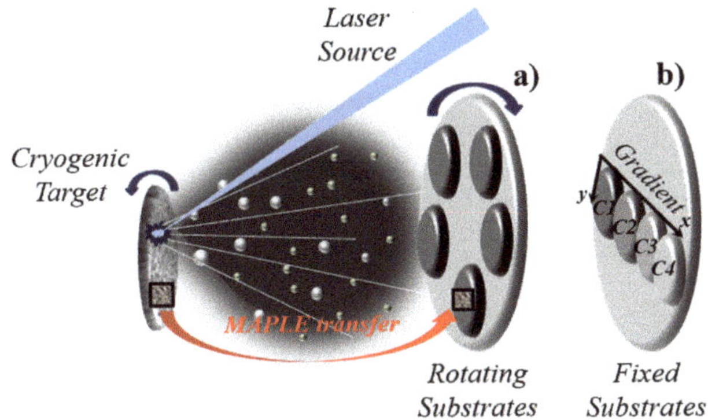

Figure 2. Schematic representation of the main deposition geometries in the case of MAPLE used to obtain coatings with uniform composition (**a**) or coatings with gradients of composition (**b**). Adapted from [65].

The experimental design employed in the case of C-MAPLE as well as the main advantages and limitations are explained in the following. In a typical irradiation geometry, two cryogenic targets are simultaneously evaporated by two pulsed laser beams. In our experimental setup, one laser beam was split and focused onto the targets. Alternatively, two distinct laser sources with different emission characteristics (wavelength, pulse duration, repetition rate) could be employed. A schematic representation of experimental setup is presented in Figure 3a. The plumes containing evaporated materials are subsequently collected on the receiving substrates, similar to MAPLE deposition. Due to plumes spreading and mixing, a gradient of composition is achieved along the longitudinal direction of the substrates, thus generating a combinatorial library as schematically depicted in Figure 3b. Fast and controlled gradients could be obtained on large areas (e.g., a glass slide scale) by adjusting the target to the substrate distance and the separation distance in-between the irradiation spots (S and D parameters in Figure 3b).

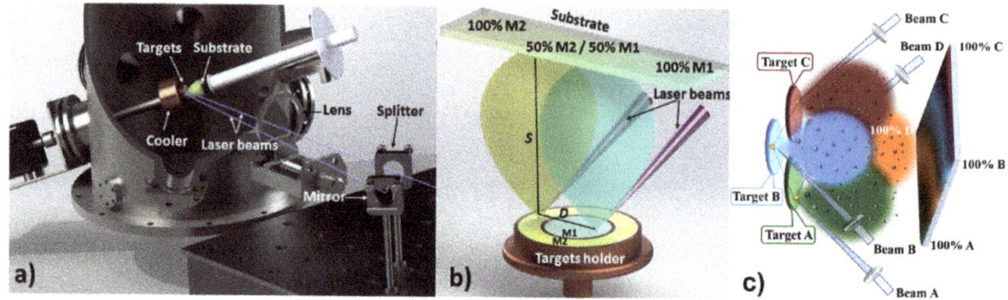

Figure 3. Experimental design of the C-MAPLE thin films processing (**a**); detailed representation of the target–substrate arrangements for combinatorial materials distribution (**b**); advanced experimental design for tailoring combinatorial diagram maps on large 2D substrates (**c**). Courtesy of M. Sopronyi (**a** and **b**) and adapted (**c**) from [63].

An advanced experimental design could be envisioned in order to generate multi-compositional diagram 2D maps on high surface areas. In this special configuration, several targets should be simultaneously irradiated using predefined laser parameters (Figure 3c). Consequently, by precisely tuning the experimental conditions, one could fabricate complex combinatorial coatings in a single-step process for high-throughput screening of the composition–structure–properties relationship at every point of the surface. Moreover, new composite materials and improved functionalities could be foreseen (e.g., drug mixtures for new therapeutic strategies).

Indeed, the basic concept of combinatorial materials science relies on synergistic mechanisms in which the interplay of two compounds with distinct properties create a novel functionality. The literature proposes various approaches to fabricate combinatorial coatings, most of them being suitable only to a limited class of compounds. Among these, we can mention several physical–chemical deposition methods such as magnetron co-sputtering deposition [141], glancing angle deposition [142], co-electrodeposition [143], casting processes [144], or flow-coating methods [145].

5.2. Ion-Doped Inorganic Bio-Coatings Obtained by Combinatorial-Matrix Assisted Pulsed Laser Evaporation

CaPs containing various amounts of different ions incorporated in the apatite crystal structure were proposed in recent research [146,147]. Among several useful divalent ions, Sr has evidenced positive effects on bone metabolism, and the introduction of Sr ranelate and use for potential treatment of osteoporosis has been demonstrated [148,149]. Zinc (Zn) also plays an essential beneficial role on reducing the risk of osteoporosis by bio-mineralization [150]. Among CaPs, β-tricalcium phosphate (β-TCP) exhibits greater solubility and resorbability compared with HA [151]. Chou et al. showed that biomimetic Zn-βTCP can stimulate faster osteogenic differentiation of MSCs to osteoblasts than pure β-TCP [152]. More recently, Boanini et al. [140] demonstrated that the C-MAPLE technique could be an alternative approach for the synthesis of gradient coatings of Sr doped HA–Zn doped β-TCP. The irradiation scheme is depicted in Figure 4a. The depositions were performed on Ti discs with a 12 mm diameter, labeled from A to E. The compositional gradient was then generated in-between A, which corresponded to 100% Sr:HA and E, which corresponded to 100% Zn:β-TCP (Figure 4a). This allowed for the demonstration of the concrete influence of the composition, structure, and topography of the implant surface on the osseointegration using an in vitro co-culture model. The preservation of the crystalline phases of the bio-coatings (with respect to the as-prepared powders used for targets) with distinct morphological features and the compositional distribution of the dopants along the gradient coatings was validated. As shown in Figure 4b, samples A and E contained Sr (green) and Zn (blue) only, while a homogenous mixture of them was naturally achieved in-between.

Human 2T-110 osteoclast precursors and human MG-63 osteoblast-like cells were co-cultured on gradient samples (arrows on Scanning Electron Microscopy - SEM images presented in Figure 4c). The response of cells was modulated by the gradual composition and strongly influenced by the Sr:HA/Zn:β-TCP ratio. In particular, Sr:HA was found to inhibit osteoclast viability and differentiation, while Zn:β-TCP proved a beneficial role on the mineralization process. The central regions of the bio-coatings exhibited rather combined effects on either osteoblast or osteoclast cells. It was concluded that gradient bio-coatings could provide materials with improved functionalities, in order to enhance and accelerate bone repair.

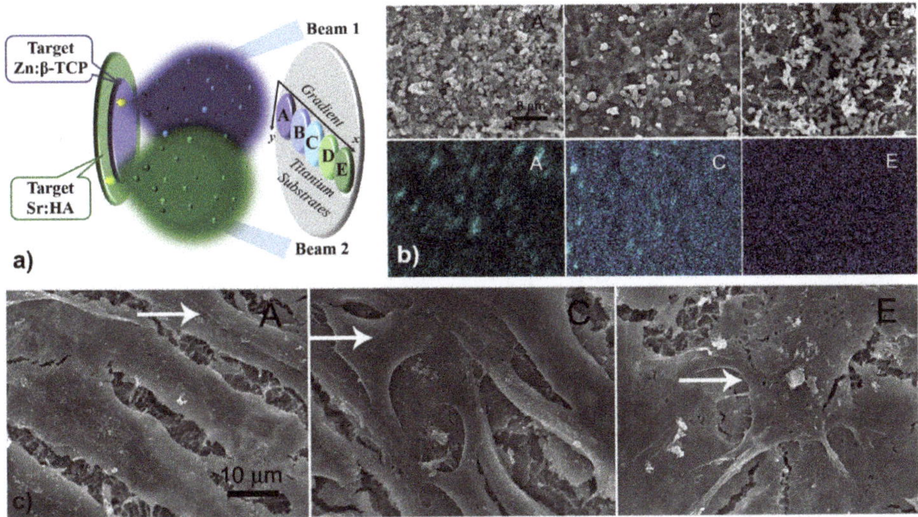

Figure 4. Experimental design for Sr:HA to Zn:β-TCP compositional gradient generation (**a**); scanning electron microscopy (SEM) images and Energy Dispersive Spectroscopy (EDS) maps of A, C and E thin films as-prepared within the compositional gradient. In the maps, the color codes represent Sr in green and Zn in blue (**b**); SEM images of osteoblasts grown on A, C and E thin films, respectively, at seven days post-seeding (**c**). Arrows in (c) indicate the bone cells. Reprinted with permission from [63,140]. (Copyright 2018 Elsevier).

5.3. Multi-Functional Organic Bio-Coatings Obtained by C-MAPLE

C-MAPLE was recently proposed for the synthesis of gradient biopolymer thin film assemblies [64] for tissue engineering and regenerative medicine applications. Synchronized laser evaporation of cryogenic targets containing sulfated *Halomonas* levan (SHL) and quaternized low molecular weight chitosan (QCH) revealed successful fabrication of combinatorial gradient bio-coatings. Studies have shown that levan polysaccharides are highly efficient in several biomedical applications such as antioxidant and anticancer activities, drug carrier systems, bioactive thin film blends, multilayer adhesive films, but also temperature responsive and cytocompatible hydrogels [153–156]. Recently, SHL has shown heparin mimetic anticoagulant activity [157] as well as improved mechanical and adhesive properties of cytocompatible and myoconductive films for cardiac tissue engineering applications [158]. Chitosan is a natural biomaterial, applied in a wide range of biomedical applications such as wound healing or tissue engineering [159,160], implant coatings [161,162], and drug delivery systems [151,163].

Mihailescu et al. [64] have demonstrated that combinatorial libraries of SHL/QCH could influence mouse fibroblasts viability (L929 cell line), coagulation effects with respect to SHL contents, and antimicrobial activity of the gradients against *Escherichia coli* and *Staphylococcus aureus* strains. The C-MAPLE design is presented in Figure 5a. Four silicon or glass substrates were positioned in front of the two evaporated plumes to be coated with composite SHL/QCH). SEM analyses of samples revealed morphological differences along the combinatorial library (Figure 5b). Specific features to each compound were observed. The irregular, interconnected polymer networks of quasi-spherical particles found in the case of SHL regions gradually disappeared with an increase in QCH content. The cell proliferation assays demonstrated good biocompatibility for all samples, with a predominance for samples consisting of 75% QCH/25% SHL (Figure 5c). This study also evidenced the highest blood clotting speed on SHL regions, which decreased with QCH content. Fluorescence microscopy images (Figure 5d) showed normal cell morphology, with polyhedral and elongated shapes, which is in agreement with the proliferation assays. The anti-biofilm activity assays revealed inhomogeneous bacterial cell adhesion at both early (8 h) and late (24 h) time points for *E. coli* along the SHL–QCH reciprocal gradient (Figure 5f). However, *S. aureus* strain tests showed that the biofilm was rather inhibited by the samples containing the highest amounts of either levan or chitosan (Figure 5e).

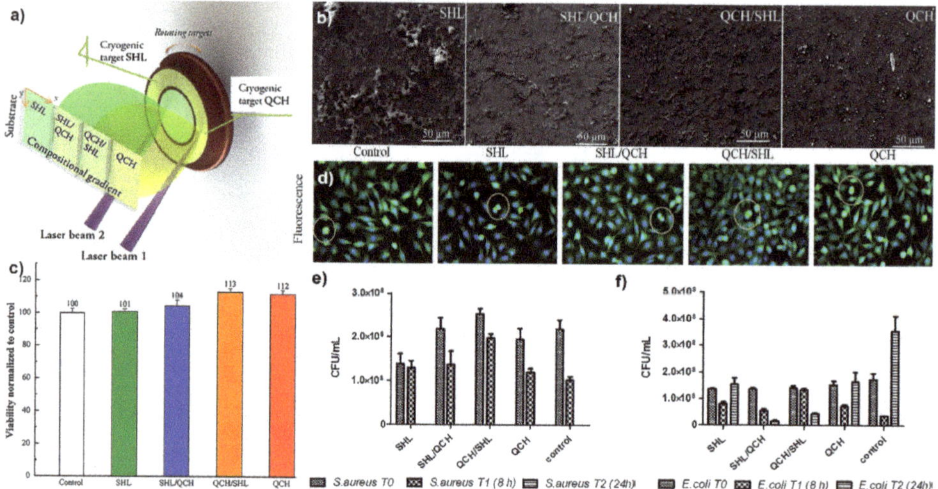

Figure 5. Experimental C-MAPLE setup (**a**); SEM micrographs of sulfated *Halomonas* levan (SHL) (**left**) to quaternized low molecular weight chitosan (QCH) (**right**) gradient bio-coatings (**b**); L929 cell proliferation on combinatorial coatings compared to the glass control (**c**). Fluorescence microscopy images of L929 cells onto the control (glass) and gradient coatings (magnification ×40). Actin fibers in green are marked with Fluorescein-Phalloidin, while nuclei in blue are stained with Hoechst. (**d**); Temporal dynamics of the *S. aureus* (**e**) and *E. coli* (**f**) biofilms growth on the C-MAPLE coatings. Circles in the fluorescence images (**d**) represent cells in various stages of division. Reprinted with permission from [64]. (Copyright 2019 Elsevier).

5.4. Hybrid Bio-Coatings by MAPLE for Anti-Tumor Drug Delivery to Cancer Cells

Very recently, Sima et al. [65] reported on the successful laser synthesis of graphene oxide nanomaterials (GON) and hybrid GON-BSA (GONB) nano-coatings incorporating anti-tumor drugs targeting melanoma cells. Among the deadliest forms of cancer, melanoma is the most aggressive skin cancer due to its high multi-drug resistance, frequent relapse, and decreased survival rates [164]. Due to their unique structural, physical, and chemical properties, carbon-based nanomaterials are extensively explored for drug/gene delivery in cancer therapy, bacteria-killing, tissue engineering platforms, engineering stem cell responses, biosensing, and cellular imaging [8]. On the other hand, nanomaterials, both synthetic and natural, may have side-effects on human tissues and generally for health, while the exposure risks are difficult to precisely evaluate [165].

In our experiments, Dabrafenib (DAB) and Trichostatin A (TSA) drugs, inhibitors for melanoma cell molecular targets, were introduced within the thin GONB coatings with successful preservation of their functional properties. DAB (chemical formula $C_{23}H_{20}F_3N_5O_2S_2$) is a drug approved by the Food and Drug Administration (FDA) in 2013, for the treatment of advanced metastatic melanoma bearing the mutated BRAF gene (BRAFV600E) found in about 70% of melanoma tumors [166]. This inhibitor proved efficient in clinical trials of phase 1 and 2 in patients with BRAFV600E mutated metastatic melanoma [167]. TSA (chemical formula $C_{17}H_{22}N_2O_3$) is a promising Histone Deacetylase (HDAC) epigenetic inhibitor used in first phase clinical trials for relapsed or refractory hematologic malignancies. HDAC inhibitors have been proven to enhance the efficacy of BRAF/MEK inhibitors in both sensitive and insensitive RAS pathway–driven melanomas [168]. A comprehensive in vitro cytotoxicity assay was conducted. For comparison, GONB solutions and coatings were tested against seven different cell lines (Table 2). It was found that for concentrations up to 37 μg/mL in water, the nanomaterials did not induce any significant decrease in cell viability for either normal or transformed cells. When exposed to the coatings, optimal cell viability was achieved on films obtained from targets that contained up to 12 μg/mL for GON and 111 μg/mL for GONB.

Table 2. Cell line characteristics.

Cells	Malignancy Phenotype	Mutation Status	Pigmentation
MNT-1	Primary melanoma	BRAF V600E	Highly pigmented
SKmel28	Primary melanoma	BRAF V600E	Amelanotic
MelJuSo	Primary melanoma	BRAF wt; N-Ras Q61K	Amelanotic
A375	Metastatic melanoma	BRAF V600E	Amelanotic
SKmel23	Metastatic melanoma	BRAF wt	Pigmented
NHEM	Normal primary melanocytes	-	Pigmented
HDF	Normal dermal fibroblasts	-	-

The determined safe concentration windows were further considered as starting points for MAPLE target preparation and deposition of the coatings containing inhibitors. The experimental design is presented in Figure 6a. A compositional gradient was generated between samples C1 and C3 in Figure 6a. SEM analysis evidenced smooth surfaces in the case of GON thin films, while the presence of BSA was found to drastically alter the morphology of the coatings, which also resulted in higher surface roughness (Figure 6b). The successful deposition and functionalization of each GONB-drug hybrid coating was further demonstrated by evaluating: (i) cellular BRAF activity inhibition and (ii) histone deacetylases activity blocking. DAB inhibition activity was validated by the decreased ERK phosphorylation in the SKmel28 primary melanoma cell line, while the TSA effect was monitored by acetylated histone accumulation in SKmel23 metastatic melanoma cell nuclei. Hence, a dose-dependent effect on target activity was evidenced for melanoma cells exposed to GONB coatings with a compositional gradient of inhibitors (Figure 6c).

Indeed, the coating efficiency was proven twice: (i) It was shown that by increasing the concentrations of GONB-TSA, an increase in fluorescence signal intensity was observed and correlated with the proportion of SKmel23 cells expressing acetyl histone H3 in the nucleus (Figure 6c left panel) and (ii) it was evidenced that by increasing the concentrations of GONB-DAB, a proportional decrease in pERK signaling in SKmel28 cells was noticed (Figure 6c, right panel) when compared to cells exposed to GONB coatings only (Figure 6c, top images). At the same time, immunofluorescence studies clearly evidenced a reduction in cell density on GONB bio-coatings loaded with a higher concentration of inhibitors (corresponding to the C1 samples).

These promising results stimulate further studies to strive for a deeper understanding of the mechanisms of laser-based coatings targeting melanoma and other cancers, with direct applications in personalized therapy studies. Such combinatorial bio-platforms could present high potential for screening cell-biomaterial interface activity for a broad spectrum of biomedical applications.

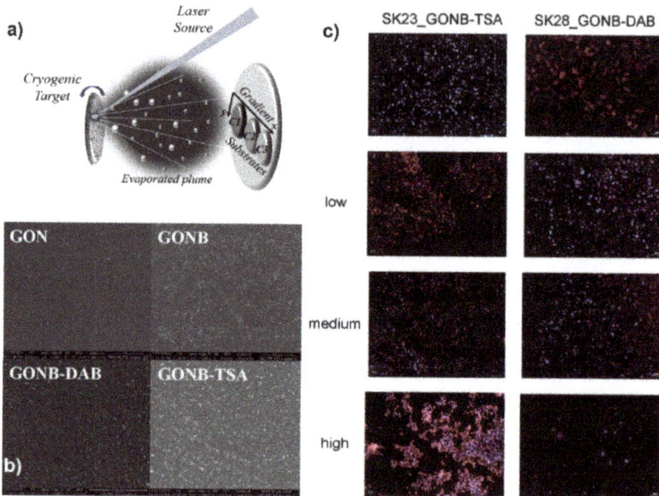

Figure 6. Experimental MAPLE setup for the synthesis of GONB-DAB/GONB-TSA composite bio-coatings with compositional gradient (fixed substrates) (**a**); SEM images of thin films deposited on silicon substrates (**b**); GONB-DAB and GONB-TSA gradients (from C3 low to C1 high) tested by immunofluorescence microscopy in comparison with simple GONB coated substrates (**c**). Here, the expressions of acetylated histone H3 (left panel) and pERK (right panel), respectively, are labeled in red, while the nuclei are labeled in blue (Hoechst). Scale bar = 50 μm. Adapted from [65].

6. Conclusions

Various technologies and methods have been applied to obtain thin bio-coatings at the interface between inert substrates and close cellular microenvironments. Most of the processes have been developed with the aim to produce either single layer, composite, or multi-layer coatings, with application specificity. Technological advancements have encouraged research to challenge biomimetic environments. Laser deposition techniques have been successfully used to fabricate bio-coatings, in particular pulsed laser deposition for inorganic materials and matrix assisted pulsed laser evaporation for both organic and inorganic materials. They have since progressed and were found applicable to either inorganic, organic, or inorganic–organic multi-layers or blended bio-coatings. Thus, combinatorial-PLD and combinatorial-MAPLE were proposed as alternatives to classic combinatorial chemical and physical deposition methods to synthesize biomimetic assemblies of complex composite and hybrid materials, with gradient of composition. C-MAPLE offers the unique characteristic of combining, in a controlled process, blended or multi-layer coating configurations of compounds dissolved in different solvent solutions, without the impediment of not being able to choose combinations with non-mixable solvents. Such combinations could allow for the synthesis of new materials with properties close to those of the native biological environment. in vitro evaluations of the bio-coatings fabricated by laser technologies confirmed their biocompatibility and capacity of modulating cell behavior. By combining the newly developed laser technologies with other chemical or physical methods, great perspectives could be open for domains like tissue engineering, nanomedicine, or controlled drug delivery in cancer research.

Funding: Partial financial support was given by the Romanian Ministry of Education and Research, under Romanian National Nucleu Program LAPLAS VI (contract No. 16N/2019) and by the Structural and Functional Proteomics Research Program of the Institute of Biochemistry of the Romanian Academy.

Acknowledgments: This review was entirely conceived, written, and processed under lockdown conditions during the COVID-19 global pandemic.We acknowledge UEFISCDI, Grant Nos: PCCDI63/2018 (PN-III-P1-1.2-PCCDI2017-0728) and TE7/2018.

Conflicts of Interest: The authors declare no conflicts of interest.

References

1. Schummer, J. Multidisciplinarity, interdisciplinarity, and patterns of research collaboration in nanoscience and nanotechnology. *Scientometrics* **2004**, *59*, 425–465. [CrossRef]
2. Zhang, F. Grand challenges for nanoscience and nanotechnology in energy and health. *Front. Chem.* **2017**, *5*, 80. [CrossRef] [PubMed]
3. Patil, A.; Mishra, V.; Thakur, S.; Riyaz, B.; Kaur, A.; Khursheed, R.; Patil, K.; Sathe, B. Nanotechnology derived nanotools in biomedical perspectives: An update. *Curr. Nanosci.* **2019**, *15*, 137–146. [CrossRef]
4. Hench, L.L.; Polak, J.M. Third-generation biomedical materials. *Science* **2002**, *295*, 1014–1017. [CrossRef] [PubMed]
5. Ning, C.; Zhou, L.; Tan, G. Fourth-generation biomedical materials. *Mater. Today* **2016**, *19*, 2–3. [CrossRef]
6. Correia, S.I.; Pereira, H.; Silva-Correia, J.; Van Dijk, C.; Espregueira-Mendes, J.; Oliveira, J.M.; Reis, R. Current concepts: Tissue engineering and regenerative medicine applications in the ankle joint. *J. R. Soc. Interface* **2014**, *11*, 20130784. [CrossRef]
7. Sima, F.; Axente, E.; Ristoscu, C.; Gallet, O.; Anselme, K.; Mihailescu, I. Bioresponsive surfaces and interfaces fabricated by innovative laser approaches. *Adv. Mater. Interfaces* **2016**, 427–462.
8. Negut, I.; Grumezescu, V.; Sima, L.E.; Axente, E. Recent advances of graphene family nanomaterials for nanomedicine. In *Fullerens, Graphenes and Nanotubes*; Elsevier: Amsterdam, The Netherlands, 2018; pp. 413–455.
9. Cohen, D.J.; Nelson, W.J.; Maharbiz, M.M. Galvanotactic control of collective cell migration in epithelial monolayers. *Nat. Mater.* **2014**, *13*, 409–417. [CrossRef]
10. Drevelle, O.; Faucheux, N. Biomimetic materials for controlling bone cell responses. *Front. Biosci.* **2013**, *5*, 369–395. [CrossRef]
11. Asri, R.; Harun, W.; Samykano, M.; Lah, N.; Ghani, S.; Tarlochan, F.; Raza, M. Corrosion and surface modification on biocompatible metals: A review. *Mater. Sci. Eng. C* **2017**, *77*, 1261–1274. [CrossRef]
12. Sridhar, T.; Mudali, U.K.; Subbaiyan, M. Preparation and characterisation of electrophoretically deposited hydroxyapatite coatings on type 316l stainless steel. *Corros. Sci.* **2003**, *45*, 237–252. [CrossRef]
13. Robin, A.; Silva, G.; Rosa, J.L. Corrosion behavior of ha-316l ss biocomposites in aqueous solutions. *Mater. Res.* **2013**, *16*, 1254–1259. [CrossRef]
14. Graziani, G.; Bianchi, M.; Sassoni, E.; Russo, A.; Marcacci, M. Ion-substituted calcium phosphate coatings deposited by plasma-assisted techniques: A review. *Mater. Sci. Eng. C* **2017**, *74*, 219–229. [CrossRef] [PubMed]
15. Xiao, D.; Zhang, J.; Zhang, C.; Barbieri, D.; Yuan, H.; Moroni, L.; Feng, G. The role of calcium phosphate surface structure in osteogenesis and the mechanism involved. *Acta Biomater.* **2020**. [CrossRef] [PubMed]
16. Zhang, J.; Dalbay, M.T.; Luo, X.; Vrij, E.; Barbieri, D.; Moroni, L.; de Bruijn, J.D.; van Blitterswijk, C.A.; Chapple, J.P.; Knight, M.M. Topography of calcium phosphate ceramics regulates primary cilia length and tgf receptor recruitment associated with osteogenesis. *Acta Biomater.* **2017**, *57*, 487–497. [CrossRef]
17. Othman, Z.; Fernandes, H.; Groot, A.J.; Luider, T.M.; Alcinesio, A.; de Melo Pereira, D.; Guttenplan, A.P.; Yuan, H.; Habibovic, P. The role of enpp1/pc-1 in osteoinduction by calcium phosphate ceramics. *Biomaterials* **2019**, *210*, 12–24. [CrossRef]
18. Bohner, M.; Miron, R.J. A proposed mechanism for material-induced heterotopic ossification. *Mater. Today* **2019**, *22*, 132–141. [CrossRef]
19. Leach, J.K.; Whitehead, J. Materials-directed differentiation of mesenchymal stem cells for tissue engineering and regeneration. *Acs Biomater. Sci. Eng.* **2017**, *4*, 1115–1127. [CrossRef]
20. Yeung, W.K.; Reilly, G.C.; Matthews, A.; Yerokhin, A. In vitro biological response of plasma electrolytically oxidized and plasma-sprayed hydroxyapatite coatings on ti–6al–4v alloy. *J. Biomed. Mater. Res. Part B Appl. Biomater.* **2013**, *101*, 939–949. [CrossRef]
21. Ke, D.; Vu, A.A.; Bandyopadhyay, A.; Bose, S. Compositionally graded doped hydroxyapatite coating on titanium using laser and plasma spray deposition for bone implants. *Acta Biomater.* **2019**, *84*, 414–423. [CrossRef]
22. Bose, S.; Fielding, G.; Tarafder, S.; Bandyopadhyay, A. Trace element doping in calcium phosphate ceramics to understand osteogenesis and angiogenesis. *Trends Biotechnol.* **2013**, *31*. [CrossRef] [PubMed]

23. Capuccini, C.; Torricelli, P.; Sima, F.; Boanini, E.; Ristoscu, C.; Bracci, B.; Socol, G.; Fini, M.; Mihailescu, I.; Bigi, A. Strontium-substituted hydroxyapatite coatings synthesized by pulsed-laser deposition: In vitro osteoblast and osteoclast response. *Acta Biomater.* **2008**, *4*, 1885–1893. [CrossRef]
24. Boanini, E.; Torricelli, P.; Fini, M.; Sima, F.; Serban, N.; Mihailescu, I.N.; Bigi, A. Magnesium and strontium doped octacalcium phosphate thin films by matrix assisted pulsed laser evaporation. *J. Inorg. Biochem.* **2012**, *107*, 65–72. [CrossRef] [PubMed]
25. Boanini, E.; Torricelli, P.; Sima, F.; Axente, E.; Fini, M.; Mihailescu, I.N.; Bigi, A. Strontium and zoledronate hydroxyapatites graded composite coatings for bone prostheses. *J. Colloid. Interface Sci.* **2015**, *448*, 1–7. [CrossRef] [PubMed]
26. Roy, M.; Bandyopadhyay, A.; Bose, S. Induction plasma sprayed sr and mg doped nano hydroxyapatite coatings on ti for bone implant. *J. Biomed. Mater. Res. Part B Appl. Biomater.* **2011**, *99*, 258–265. [CrossRef] [PubMed]
27. Tao, Z.-S.; Zhou, W.-S.; He, X.-W.; Liu, W.; Bai, B.-L.; Zhou, Q.; Huang, Z.-L.; Tu, K.-K.; Li, H.; Sun, T. A comparative study of zinc, magnesium, strontium-incorporated hydroxyapatite-coated titanium implants for osseointegration of osteopenic rats. *Mater. Sci. Eng. C* **2016**, *62*, 226–232. [CrossRef]
28. Landi, E.; Logroscino, G.; Proietti, L.; Tampieri, A.; Sandri, M.; Sprio, S. Biomimetic mg-substituted hydroxyapatite: From synthesis to in vivo behaviour. *J. Mater. Sci. Mater. Med.* **2008**, *19*, 239–247. [CrossRef]
29. Huiskes, R.; Weinans, H.; Van Rietbergen, B. The relationship between stress shielding and bone resorption around total hip stems and the effects of flexible materials. *Clin. Orthop. Relat. Res.* **1992**, 124–134. [CrossRef]
30. Scarisoreanu, N.; Craciun, F.; Ion, V.; Birjega, R.; Bercea, A.; Dinca, V.; Dinescu, M.; Sima, L.; Icriverzi, M.; Roseanu, A. Lead-free piezoelectric (ba, ca)(zr, ti) o3 thin films for biocompatible and flexible devices. *ACS Appl. Mater. Interfaces* **2017**, *9*, 266–278. [CrossRef]
31. Jannasch, M.; Gaetzner, S.; Weigel, T.; Walles, H.; Schmitz, T.; Hansmann, J. A comparative multi-parametric in vitro model identifies the power of test conditions to predict the fibrotic tendency of a biomaterial. *Sci. Rep.* **2017**, *7*, 1–17. [CrossRef]
32. Sima, L.E. Extracellular signals for guiding mesenchymal stem cells osteogenic fate. *Curr. Stem Cell Res. Ther.* **2017**, *12*, 139–144. [CrossRef] [PubMed]
33. Dhaliwal, A.; Pelka, S.; Gray, D.S.; Moghe, P.V. Engineering lineage potency and plasticity of stem cells using epigenetic molecules. *Sci. Rep.* **2018**, *8*, 1–13. [CrossRef] [PubMed]
34. Dennis, S.C.; Berkland, C.J.; Bonewald, L.F.; Detamore, M.S. Endochondral ossification for enhancing bone regeneration: Converging native extracellular matrix biomaterials and developmental engineering in vivo. *Tissue Eng. Part B Rev.* **2015**, *21*, 247–266. [CrossRef]
35. Thompson, E.M.; Matsiko, A.; Farrell, E.; Kelly, D.J.; O'Brien, F.J. Recapitulating endochondral ossification: A promising route to in vivo bone regeneration. *J. Tiss. Eng. Regen. Med.* **2015**, *9*, 889–902. [CrossRef] [PubMed]
36. Petersen, A.; Princ, A.; Korus, G.; Ellinghaus, A.; Leemhuis, H.; Herrera, A.; Klaumünzer, A.; Schreivogel, S.; Woloszyk, A.; Schmidt-Bleek, K. A biomaterial with a channel-like pore architecture induces endochondral healing of bone defects. *Nat. Commun.* **2018**, *9*, 1–16. [CrossRef] [PubMed]
37. Mikael, P.E.; Golebiowska, A.A.; Xin, X.; Rowe, D.W.; Nukavarapu, S.P. Evaluation of an engineered hybrid matrix for bone regeneration via endochondral ossification. *Ann. Biomed. Eng.* **2020**, *48*, 992–1005. [CrossRef]
38. Timashev, P.; Kuznetsova, D.; Koroleva, A.; Prodanets, N.; Deiwick, A.; Piskun, Y.; Bardakova, K.; Dzhoyashvili, N.; Kostjuk, S.; Zagaynova, E. Novel biodegradable star-shaped polylactide scaffolds for bone regeneration fabricated by two-photon polymerization. *Nanomedicine* **2016**, *11*, 1041–1053. [CrossRef]
39. Won, J.-E.; Yun, Y.-R.; Jang, J.-H.; Yang, S.-H.; Kim, J.-H.; Chrzanowski, W.; Wall, I.B.; Knowles, J.C.; Kim, H.-W. Multifunctional and stable bone mimic proteinaceous matrix for bone tissue engineering. *Biomaterials* **2015**, *56*, 46–57. [CrossRef]
40. Polini, A.; Pisignano, D.; Parodi, M.; Quarto, R.; Scaglione, S. Osteoinduction of human mesenchymal stem cells by bioactive composite scaffolds without supplemental osteogenic growth factors. *PLoS ONE* **2011**, *6*, e26211. [CrossRef]
41. Arafat, M.T.; Lam, C.X.; Ekaputra, A.K.; Wong, S.Y.; Li, X.; Gibson, I. Biomimetic composite coating on rapid prototyped scaffolds for bone tissue engineering. *Acta Biomater.* **2011**, *7*, 809–820. [CrossRef]

42. Ciapetti, G.; Granchi, D.; Devescovi, V.; Baglio, S.R.; Leonardi, E.; Martini, D.; Jurado, M.J.; Olalde, B.; Armentano, I.; Kenny, J.M. Enhancing osteoconduction of plla-based nanocomposite scaffolds for bone regeneration using different biomimetic signals to mscs. *Int. J. Mol. Sci.* **2012**, *13*, 2439–2458. [CrossRef] [PubMed]
43. Sima, F.; Axente, E.; Iordache, I.; Luculescu, C.; Gallet, O.; Anselme, K.; Mihailescu, I. Combinatorial matrix assisted pulsed laser evaporation of a biodegradable polymer and fibronectin for protein immobilization and controlled release. *Appl. Surf. Sci.* **2014**, *306*, 75–79. [CrossRef]
44. Vinson, B.T.; Phamduy, T.B.; Shipman, J.; Riggs, B.; Strong, A.L.; Sklare, S.C.; Murfee, W.L.; Burow, M.E.; Bunnell, B.A.; Huang, Y. Laser direct-write based fabrication of a spatially-defined, biomimetic construct as a potential model for breast cancer cell invasion into adipose tissue. *Biofabrication* **2017**, *9*, 025013. [CrossRef] [PubMed]
45. Coon, M.E.; Stephan, S.B.; Gupta, V.; Kealey, C.P.; Stephan, M.T. Nitinol thin films functionalized with car-t cells for the treatment of solid tumours. *Nat. Biomed. Eng.* **2020**, *4*, 195–206. [CrossRef]
46. Tobin, E.J. Recent coating developments for combination devices in orthopedic and dental applications: A literature review. *Adv. Drug Deliv. Rev.* **2017**, *112*, 88–100. [CrossRef] [PubMed]
47. Nazar, H. Developing biomaterials for tissue engineering and regenerative medicine. *Pharm. J.* **2013**, *291*, 223.
48. Aguado, B.A.; Grim, J.C.; Rosales, A.M.; Watson-Capps, J.J.; Anseth, K.S. Engineering precision biomaterials for personalized medicine. *Sci. Transl. Med.* **2018**, *10*, eaam8645. [CrossRef] [PubMed]
49. Zou, Y.; Zhang, L.; Yang, L.; Zhu, F.; Ding, M.; Lin, F.; Wang, Z.; Li, Y. "Click" chemistry in polymeric scaffolds: Bioactive materials for tissue engineering. *J. Control. Release* **2018**, *273*, 160–179. [CrossRef]
50. Surmenev, R.A.; Surmeneva, M.A. A critical review of decades of research on calcium-phosphate-based coatings: How far are we from their widespread clinical application? *Curr. Opin. Biomed. Eng.* **2019**. [CrossRef]
51. Surmenev, R.A.; Surmeneva, M.A.; Ivanova, A.A. Significance of calcium phosphate coatings for the enhancement of new bone osteogenesis–a review. *Acta Biomater.* **2014**, *10*, 557–579. [CrossRef]
52. Van Hoof, M.; Wigren, S.; Ivarsson Blechert, J.; Molin, M.; Andersson, H.; Mateijsen, D.; Bom, S.J.; Calmels, M.; van der Rijt, A.J.; Flynn, M.C. A multinational cost-consequence analysis of a bone conduction hearing implant system—a randomized trial of a conventional vs. A less invasive treatment with new abutment technology. *Front. Neurol.* **2020**, *11*, 106. [CrossRef] [PubMed]
53. Wikesjö, U.M.; Susin, C.; Qahash, M.; Polimeni, G.; Leknes, K.N.; Shanaman, R.H.; Prasad, H.S.; Rohrer, M.D.; Hall, J. The critical-size supraalveolar peri-implant defect model: Characteristics and use. *J. Clin. Periodontol.* **2006**, *33*, 846–854. [CrossRef] [PubMed]
54. León, B.; Jansen, J.A. Thin Calcium Phosphate Coatings for Medical Implants. Available online: http://public.eblib.com/EBLPublic/PublicView.do?ptiID=450789 (accessed on 9 May 2020).
55. Popescu-Pelin, G.; Sima, F.; Sima, L.; Mihailescu, C.; Luculescu, C.; Iordache, I.; Socol, M.; Socol, G.; Mihailescu, I. Hydroxyapatite thin films grown by pulsed laser deposition and matrix assisted pulsed laser evaporation: Comparative study. *Appl. Surf. Sci.* **2017**, *418*, 580–588. [CrossRef]
56. Shaikh, S.; Kedia, S.; Majumdar, A.G.; Subramanian, M.; Sinha, S. 45s5 bioactive glass coating on ti6al4v alloy using pulsed laser deposition technique. *Mater. Res. Express* **2020**, *6*, 125428. [CrossRef]
57. Socol, G.; Socol, M.; Sima, L.; Petrescu, S.; Enculescu, M.; Sima, F.; Miroiu, M.; Popescu-Pelin, G.; Stefan, N.; Cristescu, R. Combinatorial pulsed laser deposition of ag-containing calcium phosphate coatings. *Dig. J. Nanomat. Biostruct.* **2012**, *7*, 563–576.
58. Mihailescu, I.N.; Bociaga, D.; Socol, G.; Stan, G.E.; Chifiriuc, M.-C.; Bleotu, C.; Husanu, M.A.; Popescu-Pelin, G.; Duta, L.; Luculescu, C.R. Fabrication of antimicrobial silver-doped carbon structures by combinatorial pulsed laser deposition. *Int. J. Pharm.* **2016**, *515*, 592–606. [CrossRef]
59. Icriverzi, M.; Rusen, L.; Brajnicov, S.; Bonciu, A.; Dinescu, M.; Cimpean, A.; Evans, R.W.; Dinca, V.; Roseanu, A. Macrophage in vitro response on hybrid coatings obtained by matrix assisted pulsed laser evaporation. *Coatings* **2019**, *9*, 236. [CrossRef]
60. Negut, I.; Floroian, L.; Ristoscu, C.; Mihailescu, C.N.; Mirza Rosca, J.C.; Tozar, T.; Badea, M.; Grumezescu, V.; Hapenciuc, C.; Mihailescu, I.N. Functional bioglass—biopolymer double nanostructure for natural antimicrobial drug extracts delivery. *Nanomaterials* **2020**, *10*, 385. [CrossRef]
61. Visan, A.; Stan, G.E.; Ristoscu, C.; Popescu-Pelin, G.; Sopronyi, M.; Besleaga, C.; Luculescu, C.; Chifiriuc, M.C.; Hussien, M.; Marsan, O. Combinatorial maple deposition of antimicrobial orthopedic maps fabricated from chitosan and biomimetic apatite powders. *Int. J. Pharm.* **2016**, *511*, 505–515. [CrossRef]

62. Axente, E.; Ristoscu, C.; Bigi, A.; Sima, F.; Mihailescu, I.N. Combinatorial laser synthesis of biomaterial thin films: Selection and processing for medical applications. In *Advances in the Application of Lasers in Materials Science*; Springer: Berlin, Germany, 2018; pp. 309–338.
63. Axente, E.; Sima, F. Biomimetic nanostructures with compositional gradient grown by combinatorial matrix-assisted pulsed laser evaporation for tissue engineering. *Curr. Med. Chem.* **2020**, *27*, 903–918. [CrossRef]
64. Mihailescu, N.; Haskoylu, M.E.; Ristoscu, C.; Bostan, M.S.; Sopronyi, M.; Eroğlu, M.S.; Chifiriuc, M.C.; Mustaciosu, C.C.; Axente, E.; Oner, E.T. Gradient multifunctional biopolymer thin film assemblies synthesized by combinatorial maple. *Appl. Surf. Sci.* **2019**, *466*, 628–636. [CrossRef]
65. Sima, L.E.; Chiritoiu, G.; Negut, I.; Grumezescu, V.; Orobeti, S.; Munteanu, C.V.; Sima, F.; Axente, E. Functionalized graphene oxide thin films for anti-tumor drug delivery to melanoma cells. *Front. Chem.* **2020**, *8*. [CrossRef]
66. Hakobyan, D.; Kerouredan, O.; Remy, M.; Dusserre, N.; Medina, C.; Devillard, R.; Fricain, J.-C.; Oliveira, H. Laser-assisted bioprinting for bone repair. In *3d Bioprinting*; Springer: Berlin, Germany, 2020; pp. 135–144.
67. Papavlu, A.P.; Dinca, V.; Dinescu, M. Laser printing of proteins and biomaterials. *Laser Print. Funct. Mater. 3d Microfabr. Electron. Biomed.* **2018**.
68. Keriquel, V.; Oliveira, H.; Rémy, M.; Ziane, S.; Delmond, S.; Rousseau, B.; Rey, S.; Catros, S.; Amédée, J.; Guillemot, F. In situ printing of mesenchymal stromal cells, by laser-assisted bioprinting, for in vivo bone regeneration applications. *Sci. Rep.* **2017**, *7*, 1–10. [CrossRef]
69. Pichugin, V.F.; Surmenev, R.A.; Shesterikov, E.V.; Ryabtseva, M.; Eshenko, E.; Tverdokhlebov, S.I.; Prymak, O.; Epple, M. The preparation of calcium phosphate coatings on titanium and nickel–titanium by rf-magnetron-sputtered deposition: Composition, structure and micromechanical properties. *Surf. Coat. Technol.* **2008**, *202*, 3913–3920. [CrossRef]
70. Fedotkin, A.; Bolbasov, E.; Kozelskaya, A.; Dubinenko, G.; Shesterikov, E.; Ashrafov, A.; Tverdokhlebov, S. Calcium phosphate coating deposition by radio frequency magnetron sputtering in the various inert gases: The pilot study. *Mater. Chem. Phys.* **2019**, *235*, 121735. [CrossRef]
71. Stan, G.E.; Pasuk, I.; Husanu, M.A.; Enculescu, I.; Pina, S.; Lemos, A.F.; Tulyaganov, D.U.; El Mabrouk, K.; Ferreira, J.M. Highly adherent bioactive glass thin films synthetized by magnetron sputtering at low temperature. *J. Mater. Sci. Mater. Med.* **2011**, *22*, 2693–2710. [CrossRef] [PubMed]
72. Sima, L.E.; Stan, G.E.; Morosanu, C.O.; Melinescu, A.; Ianculescu, A.; Melinte, R.; Neamtu, J.; Petrescu, S.M. Differentiation of mesenchymal stem cells onto highly adherent radio frequency-sputtered carbonated hydroxylapatite thin films. *J. Biomed. Mater. Res. Part A* **2010**, *95*, 1203–1214. [CrossRef] [PubMed]
73. Popa, A.; Stan, G.; Besleaga, C.; Ion, L.; Maraloiu, V.; Tulyaganov, D.; Ferreira, J. Submicrometer hollow bioglass cones deposited by radio frequency magnetron sputtering: Formation mechanism, properties, and prospective biomedical applications. *ACS Appl. Mater. Interfaces* **2016**, *8*, 4357–4367. [CrossRef] [PubMed]
74. Liguori, A.; Gualandi, C.; Focarete, M.L.; Biscarini, F.; Bianchi, M. The pulsed electron deposition technique for biomedical applications: A review. *Coatings* **2020**, *10*, 16. [CrossRef]
75. Unabia, R.B.; Bonebeau, S.; Candidato, R.T., Jr.; Pawłowski, L. Preliminary study on copper-doped hydroxyapatite coatings obtained using solution precursor plasma spray process. *Surf. Coat. Technol.* **2018**, *353*, 370–377. [CrossRef]
76. Vilardell, A.; Cinca, N.; Garcia-Giralt, N.; Dosta, S.; Cano, I.; Nogués, X.; Guilemany, J. In-vitro comparison of hydroxyapatite coatings obtained by cold spray and conventional thermal spray technologies. *Mater. Sci. Eng. C* **2020**, *107*, 110306. [CrossRef] [PubMed]
77. Henao, J.; Poblano-Salas, C.; Monsalve, M.; Corona-Castuera, J.; Barceinas-Sanchez, O. Bio-active glass coatings manufactured by thermal spray: A status report. *J. Mater. Res. Technol.* **2019**. [CrossRef]
78. Alkekhia, D.; Hammond, P.T.; Shukla, A. Layer-by-layer biomaterials for drug delivery. *Annu. Rev. Biomed. Eng.* **2020**, *22*. [CrossRef] [PubMed]
79. Visan, A.I.; Popescu-Pelin, G.; Gherasim, O.; Mihailescu, A.; Socol, M.; Zgura, I.; Chiritoiu, M.; Elena Sima, L.; Antohe, F.; Ivan, L. Long-term evaluation of dip-coated pcl-blend-peg coatings in simulated conditions. *Polymers* **2020**, *12*, 717. [CrossRef] [PubMed]
80. Brigaud, I.; Agniel, R.; Leroy-Dudal, J.; Kellouche, S.; Ponche, A.; Bouceba, T.; Mihailescu, N.; Sopronyi, M.; Viguier, E.; Ristoscu, C. Synergistic effects of bmp-2, bmp-6 or bmp-7 with human plasma fibronectin onto hydroxyapatite coatings: A comparative study. *Acta Biomater.* **2017**, *55*, 481–492. [CrossRef]

81. Banta, R.A.; Collins, T.W.; Curley, R.; O'Connell, J.; Young, P.W.; Holmes, J.D.; Flynn, E.J. Regulated phase separation in nanopatterned protein-polysaccharide thin films by spin coating. *Colloids Surf. B Biointerfaces* **2020**, *190*, 110967. [CrossRef]
82. Ishikawa, K.; Garskaite, E.; Kareiva, A. Sol–gel synthesis of calcium phosphate-based biomaterials—A review of environmentally benign, simple, and effective synthesis routes. *J. Sol Gel Sci. Technol.* **2020**, 1–22. [CrossRef]
83. Ballarre, J.; Aydemir, T.; Liverani, L.; Roether, J.; Goldmann, W.; Boccaccini, A. Versatile bioactive and antibacterial coating system based on silica, gentamicin, and chitosan: Improving early stage performance of titanium implants. *Surf. Coat. Technol.* **2020**, *381*, 125138. [CrossRef]
84. Boccaccini, A.R.; Keim, S.; Ma, R.; Li, Y.; Zhitomirsky, I. Electrophoretic deposition of biomaterials. *J. R. Soc. Interface R. Soc.* **2010**, *7* (Suppl. 5), S581–S613. [CrossRef]
85. Singh, S.; Singh, G.; Bala, N. Electrophoretic deposition of hydroxyapatite-iron oxide-chitosan composite coatings on ti-13nb-13zr alloy for biomedical applications. *Thin Solid Film.* **2020**, *697*, 137801. [CrossRef]
86. Lu, Y.-P.; Li, M.-S.; Li, S.-T.; Wang, Z.-G.; Zhu, R.-F. Plasma-sprayed hydroxyapatite+ titania composite bond coat for hydroxyapatite coating on titanium substrate. *Biomaterials* **2004**, *25*, 4393–4403. [CrossRef] [PubMed]
87. Grubova, I.Y.; Surmeneva, M.A.; Ivanova, A.A.; Kravchuk, K.; Prymak, O.; Epple, M.; Buck, V.; Surmenev, R.A. The effect of patterned titanium substrates on the properties of silver-doped hydroxyapatite coatings. *Surf. Coat. Technol.* **2015**, *276*, 595–601. [CrossRef]
88. Nelea, V.; Ristoscu, C.; Chiritescu, C.; Ghica, C.; Mihailescu, I.; Pelletier, H.; Mille, P.; Cornet, A. Pulsed laser deposition of hydroxyapatite thin films on ti-5al-2.5 fe substrates with and without buffer layers. *Appl. Surf. Sci.* **2000**, *168*, 127–131. [CrossRef]
89. Nelea, V.; Morosanu, C.; Iliescu, M.; Mihailescu, I. Hydroxyapatite thin films grown by pulsed laser deposition and radio-frequency magnetron sputtering: Comparative study. *Appl. Surf. Sci.* **2004**, *228*, 346–356. [CrossRef]
90. Duta, L.; Popescu, A.C. Current status on pulsed laser deposition of coatings from animal-origin calcium phosphate sources. *Coatings* **2019**, *9*, 335. [CrossRef]
91. Garcia-Sanz, F.; Mayor, M.; Arias, J.; Pou, J.; Leon, B.; Perez-Amor, M. Hydroxyapatite coatings: A comparative study between plasma-spray and pulsed laser deposition techniques. *J. Mater. Sci. Mater. Med.* **1997**, *8*, 861–865. [CrossRef]
92. Mohseni, E.; Zalnezhad, E.; Bushroa, A.R. Comparative investigation on the adhesion of hydroxyapatite coating on ti–6al–4v implant: A review paper. *Int. J. Adhes. Adhes.* **2014**, *48*, 238–257. [CrossRef]
93. Smith, J.R.; Lamprou, D.A. Polymer coatings for biomedical applications: A review. *Trans. Imf.* **2014**, *92*, 9–19. [CrossRef]
94. Hallab, N.J.; Bundy, K.J.; O'Connor, K.; Moses, R.L.; Jacobs, J.J. Evaluation of metallic and polymeric biomaterial surface energy and surface roughness characteristics for directed cell adhesion. *Tissue Eng.* **2001**, *7*, 55–71. [CrossRef]
95. Mihailescu, I.N.; Bigi, A.; Gyorgy, E.; Ristoscu, C.; Sima, F.; Oner, E.T. Biomaterial thin films by soft pulsed laser technologies for biomedical applications. In *Lasers in Materials Science*; Springer: Berlin, Germany, 2014; pp. 271–294.
96. Sima, F.; Davidson, P.M.; Dentzer, J.; Gadiou, R.; Pauthe, E.; Gallet, O.; Mihailescu, I.N.; Anselme, K. Inorganic–organic thin implant coatings deposited by lasers. *ACS Appl. Mater. Interfaces* **2015**, *7*, 911–920. [CrossRef] [PubMed]
97. Axente, E.; Sima, F.; Ristoscu, C.; Mihailescu, N.; Mihailescu, I.N. Biopolymer thin films synthesized by advanced pulsed laser techniques. *Recent Adv. Biopolym.* **2016**, *73*.
98. Stiff-Roberts, A.D.; Ge, W. Organic/hybrid thin films deposited by matrix-assisted pulsed laser evaporation (maple). *Appl. Phys. Rev.* **2017**, *4*, 041303. [CrossRef]
99. Schmidt, V.; Belegratis, M.R. Introduction and scope of the book. In *Laser Technology in Biomimetics*; Springer: Berlin, Germany, 2013; pp. 1–12.
100. Craciun, D.; Craciun, V. Reactive pulsed laser deposition of tin. *Appl. Surf. Sci.* **1992**, *54*, 75–77. [CrossRef]
101. Lunney, J.G. Pulsed laser deposition of metal and metal multilayer films. *Appl. Surf. Sci.* **1995**, *86*, 79–85. [CrossRef]

102. Caricato, A.P.; Luches, A. Applications of the matrix-assisted pulsed laser evaporation method for the deposition of organic, biological and nanoparticle thin films: A review. *Appl. Phys. A* **2011**, *105*, 565–582. [CrossRef]
103. Sima, F.; Mutlu, E.C.; Eroglu, M.S.; Sima, L.E.; Serban, N.; Ristoscu, C.; Petrescu, S.M.; Oner, E.T.; Mihailescu, I.N. Levan nanostructured thin films by maple assembling. *Biomacromolecules* **2011**, *12*, 2251–2256. [CrossRef]
104. Motoc, M.; Axente, E.; Popescu, C.; Sima, L.; Petrescu, S.; Mihailescu, I.; Gyorgy, E. Active protein and calcium hydroxyapatite bilayers grown by laser techniques for therapeutic applications. *J. Biomed. Mater. Res. Part A* **2013**, *101*, 2706–2711. [CrossRef]
105. Dinca, V.; Florian, P.E.; Sima, L.E.; Rusen, L.; Constantinescu, C.; Evans, R.W.; Dinescu, M.; Roseanu, A. Maple-based method to obtain biodegradable hybrid polymeric thin films with embedded antitumoral agents. *Biomed. Microdev.* **2014**, *16*, 11–21. [CrossRef]
106. Sima, F.; Davidson, P.; Pauthe, E.; Sima, L.; Gallet, O.; Mihailescu, I.; Anselme, K. Fibronectin layers by matrix-assisted pulsed laser evaporation from saline buffer-based cryogenic targets. *Acta Biomater.* **2011**, *7*, 3780–3788. [CrossRef]
107. Sima, F.; Davidson, P.; Pauthe, E.; Gallet, O.; Anselme, K.; Mihailescu, I. Thin films of vitronectin transferred by maple. *Appl. Phys. A* **2011**, *105*, 611–617. [CrossRef]
108. McGill, R.A.; Chrisey, D.B. Method of Producing a Film Coating by Matrix Assisted Pulsed Laser Deposition. U.S. Patent 6,025,036, 15 February 2000.
109. Jelinek, M.; Kocourek, T.; Remsa, J.; Cristescu, R.; Mihailescu, I.; Chrisey, D. Maple applications in studying organic thin films. *Laser Phys.* **2007**, *17*, 66–70. [CrossRef]
110. Califano, V.; Bloisi, F.; Vicari, L.R.; Colombi, P.; Bontempi, E.; Depero, L.E. Maple deposition of biomaterial multilayers. *Appl. Surf. Sci.* **2008**, *254*, 7143–7148. [CrossRef]
111. Stamatin, L.; Cristescu, R.; Socol, G.; Moldovan, A.; Mihaiescu, D.; Stamatin, I.; Mihailescu, I.; Chrisey, D. Laser deposition of fibrinogen blood proteins thin films by matrix assisted pulsed laser evaporation. *Appl. Surf. Sci.* **2005**, *248*, 422–427. [CrossRef]
112. Guha, S.; Adil, D.; Ukah, N.; Gupta, R.; Ghosh, K. Maple-deposited polymer films for improved organic device performance. *Appl. Phys. A* **2011**, *105*, 547–554. [CrossRef]
113. Axente, E.; Sopronyi, M.; Ghimbeu, C.M.; Nita, C.; Airoudj, A.; Schrodj, G.; Sima, F. Matrix-assisted pulsed laser evaporation: A novel approach to design mesoporous carbon films. *Carbon* **2017**, *122*, 484–495. [CrossRef]
114. Del Pino, Á.P.; Ramadan, M.A.; Lebière, P.G.; Ivan, R.; Logofatu, C.; Yousef, I.; György, E. Fabrication of graphene-based electrochemical capacitors through reactive inverse matrix assisted pulsed laser evaporation. *Appl. Surf. Sci.* **2019**, *484*, 245–256. [CrossRef]
115. Del Pino, Á.P.; López, M.R.; Ramadan, M.A.; Lebière, P.G.; Logofatu, C.; Martínez-Rovira, I.; Yousef, I.; György, E. Enhancement of the supercapacitive properties of laser deposited graphene-based electrodes through carbon nanotube loading and nitrogen doping. *Phys. Chem. Chem. Phys.* **2019**, *21*, 25175–25186. [CrossRef]
116. Palla-Papavlu, A.; Constantinescu, C.; Dinca, V.; Matei, A.; Moldovan, A.; Mitu, B.; Dinescu, M. Polyisobutylene thin films obtained by matrix assisted pulsed laser evaporation for sensors applications. *Sens. Lett.* **2010**, *8*, 502–506. [CrossRef]
117. Palla-Papavlu, A.; Dinca, V.; Dinescu, M.; Di Pietrantonio, F.; Cannatà, D.; Benetti, M.; Verona, E. Matrix-assisted pulsed laser evaporation of chemoselective polymers. *Appl. Phys. A* **2011**, *105*, 651–659. [CrossRef]
118. Dinca, V.; Viespe, C.; Brajnicov, S.; Constantinoiu, I.; Moldovan, A.; Bonciu, A.; Toader, C.N.; Ginghina, R.E.; Grigoriu, N.; Dinescu, M. Maple assembled acetylcholinesterase–polyethylenimine hybrid and multilayered interfaces for toxic gases detection. *Sensors* **2018**, *18*, 4265. [CrossRef] [PubMed]
119. Caricato, A.P.; Ge, W.; Stiff-Roberts, A.D. Uv-and rir-maple: Fundamentals and applications. In *Advances in the Application of Lasers in Materials Science*; Springer: Berlin/Heidelberg, Germany, 2018; pp. 275–308.
120. Smausz, T.; Megyeri, G.; Kékesi, R.; Vass, C.; György, E.; Sima, F.; Mihailescu, I.N.; Hopp, B. Comparative study on pulsed laser deposition and matrix assisted pulsed laser evaporation of urease thin films. *Thin Solid Film* **2009**, *517*, 4299–4302. [CrossRef]

121. György, E.; Sima, F.; Mihailescu, I.; Smausz, T.; Megyeri, G.; Kékesi, R.; Hopp, B.; Zdrentu, L.; Petrescu, S. Immobilization of urease by laser techniques: Synthesis and application to urea biosensors. *J. Biomed. Mater. Res. Part A Off. J. Soc. Biomater. Jpn. Soc. Biomater. Aust. Soc. Biomater. Korean Soc. Biomater.* **2009**, *89*, 186–191. [CrossRef] [PubMed]
122. Predoi, D.; Ciobanu, C.S.; Radu, M.; Costache, M.; Dinischiotu, A.; Popescu, C.; Axente, E.; Mihailescu, I.; Gyorgy, E. Hybrid dextran-iron oxide thin films deposited by laser techniques for biomedical applications. *Mater. Sci. Eng. C* **2012**, *32*, 296–302. [CrossRef]
123. Axente, E.; Sima, F.; Sima, L.E.; Erginer, M.; Eroglu, M.S.; Serban, N.; Ristoscu, C.; Petrescu, S.M.; Oner, E.T.; Mihailescu, I.N. Combinatorial maple gradient thin film assemblies signalling to human osteoblasts. *Biofabrication* **2014**, *6*, 035010. [CrossRef] [PubMed]
124. Miroiu, F.; Socol, G.; Visan, A.; Stefan, N.; Craciun, D.; Craciun, V.; Dorcioman, G.; Mihailescu, I.; Sima, L.; Petrescu, S. Composite biocompatible hydroxyapatite–silk fibroin coatings for medical implants obtained by matrix assisted pulsed laser evaporation. *Mater. Sci. Eng. B* **2010**, *169*, 151–158. [CrossRef]
125. Rașoga, O.; Sima, L.; Chirițoiu, M.; Popescu-Pelin, G.; Fufă, O.; Grumezescu, V.; Socol, M.; Stănculescu, A.; Zgură, I.; Socol, G. Biocomposite coatings based on poly (3-hydroxybutyrate-co-3-hydroxyvalerate)/calcium phosphates obtained by maple for bone tissue engineering. *Appl. Surf. Sci.* **2017**, *417*, 204–212. [CrossRef]
126. Cristescu, R.; Popescu, C.; Socol, G.; Iordache, I.; Mihailescu, I.; Mihaiescu, D.; Grumezescu, A.; Balan, A.; Stamatin, I.; Chifiriuc, C. Magnetic core/shell nanoparticle thin films deposited by maple: Investigation by chemical, morphological and in vitro biological assays. *Appl. Surf. Sci.* **2012**, *258*, 9250–9255. [CrossRef]
127. Grumezescu, A.M.; Cristescu, R.; Chifiriuc, M.; Dorcioman, G.; Socol, G.; Mihailescu, I.; Mihaiescu, D.E.; Ficai, A.; Vasile, O.; Enculescu, M. Fabrication of magnetite-based core–shell coated nanoparticles with antibacterial properties. *Biofabrication* **2015**, *7*, 015014. [CrossRef]
128. Dinca, V.; Liu, Q.; Brajnicov, S.; Bonciu, A.; Vlad, A.; Dinu, C.Z. Composites formed from tungsten trioxide and graphene oxide for the next generation of electrochromic interfaces. *Compos. Commun.* **2020**, *17*, 115–122. [CrossRef]
129. Greer, J.A. Design challenges for matrix assisted pulsed laser evaporation and infrared resonant laser evaporation equipment. *Appl. Phys. A* **2011**, *105*, 661–671. [CrossRef]
130. Bubb, D.; O'Malley, S.; Antonacci, C.; Simonson, D.; McGill, R. Matrix-assisted laser deposition of a sorbent oligomer using an infrared laser. *J. Appl. Phys.* **2004**, *95*, 2175–2177. [CrossRef]
131. Pate, R.; McCormick, R.; Chen, L.; Zhou, W.; Stiff-Roberts, A.D. Rir-maple deposition of conjugated polymers for application to optoelectronic devices. *Appl. Phys. A* **2011**, *105*, 555–563. [CrossRef]
132. Piqué, A. The matrix-assisted pulsed laser evaporation (maple) process: Origins and future directions. *Appl. Phys. A* **2011**, *105*, 517–528. [CrossRef]
133. Zhigilei, L.V.; Leveugle, E.; Garrison, B.J.; Yingling, Y.G.; Zeifman, M.I. Computer simulations of laser ablation of molecular substrates. *Chem. Rev.* **2003**, *103*, 321–348. [CrossRef] [PubMed]
134. Leveugle, E.; Zhigilei, L.V. Molecular dynamics simulation study of the ejection and transport of polymer molecules in matrix-assisted pulsed laser evaporation. *J. Appl. Phys.* **2007**, *102*, 074914. [CrossRef]
135. Tabetah, M.; Matei, A.; Constantinescu, C.; Mortensen, N.P.; Dinescu, M.; Schou, J.; Zhigilei, L.V. The minimum amount of "matrix" needed for matrix-assisted pulsed laser deposition of biomolecules. *J. Phys. Chem. B* **2014**, *118*, 13290–13299. [CrossRef]
136. Yingling, Y.G.; Garrison, B.J. Coarse-grained chemical reaction model. *J. Phys. Chem. B* **2004**, *108*, 1815–1821. [CrossRef]
137. Torres, R.D.; Johnson, S.L.; Haglund, R.F.; Hwang, J.; Burn, P.L.; Holloway, P.H. Mechanisms of resonant infrared matrix-assisted pulsed laser evaporation. *Crit. Rev. Solid State Mater. Sci.* **2011**, *36*, 16–45. [CrossRef]
138. Sima, F.; Axente, E.; Sima, L.E.; Tuyel, U.; Eroglu, M.S.; Serban, N.; Ristoscu, C.; Petrescu, S.M.; Oner, E.T.; Mihailescu, I.N. Combinatorial matrix-assisted pulsed laser evaporation: Single-step synthesis of biopolymer compositional gradient thin film assemblies. *Appl. Phys. Lett.* **2012**, *101*, 233705. [CrossRef]
139. Takeuchi, I. Combinatorial pulsed laser deposition. In *Pulsed Laser Deposition of Thin Films: Applications-Led Growth of Functional Materials*; John Wiley Sons Inc.: Hoboken, NJ, USA, 2007; pp. 161–175.
140. Boanini, E.; Torricelli, P.; Sima, F.; Axente, E.; Fini, M.; Mihailescu, I.N.; Bigi, A. Gradient coatings of strontium hydroxyapatite/zinc β-tricalcium phosphate as a tool to modulate osteoblast/osteoclast response. *J. Inorg. Biochem.* **2018**, *183*, 1–8. [CrossRef] [PubMed]

141. Xing, H.; Zhao, B.; Wang, Y.; Zhang, X.; Ren, Y.; Yan, N.; Gao, T.; Li, J.; Zhang, L.; Wang, H. Rapid construction of fe–co–ni composition-phase map by combinatorial materials chip approach. *ACS Comb. Sci.* **2018**, *20*, 127–131. [CrossRef] [PubMed]
142. Motemani, Y.; Khare, C.; Savan, A.; Hans, M.; Paulsen, A.; Frenzel, J.; Somsen, C.; Mücklich, F.; Eggeler, G.; Ludwig, A. Nanostructured ti–ta thin films synthesized by combinatorial glancing angle sputter deposition. *Nanotechnology* **2016**, *27*, 495604. [CrossRef] [PubMed]
143. Sun, F.; Sask, K.; Brash, J.; Zhitomirsky, I. Surface modifications of nitinol for biomedical applications. *Colloids Surf. B Biointerfaces* **2008**, *67*, 132–139. [CrossRef] [PubMed]
144. Hoogenboom, R.; Meier, M.A.; Schubert, U.S. Combinatorial methods, automated synthesis and high-throughput screening in polymer research: Past and present. *Macromol. Rapid Commun.* **2003**, *24*, 15–32. [CrossRef]
145. Meredith, J.C.; Smith, A.P.; Karim, A.; Amis, E.J. Combinatorial materials science for polymer thin-film dewetting. *Macromolecules* **2000**, *33*, 9747–9756. [CrossRef]
146. Boanini, E.; Gazzano, M.; Bigi, A. Ionic substitutions in calcium phosphates synthesized at low temperature. *Acta Biomater.* **2010**, *6*, 1882–1894. [CrossRef]
147. Arcos, D.; Vallet-Regí, M. Substituted hydroxyapatite coatings of bone implants. *J. Mater. Chem. B* **2020**, *8*, 1781–1800. [CrossRef]
148. Marie, P. Strontium ranelate: A novel mode of action optimizing bone formation and resorption. *Osteoporos. Int.* **2005**, *16*, S7–S10. [CrossRef]
149. Gallacher, S.; Dixon, T. Impact of treatments for postmenopausal osteoporosis (bisphosphonates, parathyroid hormone, strontium ranelate, and denosumab) on bone quality: A systematic review. *Calcif. Tissue Int.* **2010**, *87*, 469–484. [CrossRef]
150. Lynch, R.J. Zinc in the mouth, its interactions with dental enamel and possible effects on caries; a review of the literature. *Int. Dent. J.* **2011**, *61*, 46–54. [CrossRef] [PubMed]
151. Bostan, M.S.; Senol, M.; Cig, T.; Peker, I.; Goren, A.C.; Ozturk, T.; Eroglu, M.S. Controlled release of 5-aminosalicylicacid from chitosan based ph and temperature sensitive hydrogels. *Int. J. Biol. Macromol.* **2013**, *52*, 177–183. [CrossRef] [PubMed]
152. Chou, J.; Hao, J.; Hatoyama, H.; Ben-Nissan, B.; Milthorpe, B.; Otsuka, M. Effect of biomimetic zinc-containing tricalcium phosphate (zn–tcp) on the growth and osteogenic differentiation of mesenchymal stem cells. *J. Tissue Eng. Regen. Med.* **2015**, *9*, 852–858. [CrossRef] [PubMed]
153. Öner, E.T.; Hernández, L.; Combie, J. Review of levan polysaccharide: From a century of past experiences to future prospects. *Biotechnol. Adv.* **2016**, *34*, 827–844. [CrossRef]
154. Berg, A.; Oner, E.T.; Combie, J.; Schneider, B.; Ellinger, R.; Weisser, J.; Wyrwa, R.; Schnabelrauch, M. Formation of new, cytocompatible hydrogels based on photochemically crosslinkable levan methacrylates. *Int. J. Biol. Macromol.* **2018**, *107*, 2312–2319. [CrossRef]
155. Demirci, T.; Hasköylü, M.E.; Eroğlu, M.S.; Hemberger, J.; Öner, E.T. Levan-based hydrogels for controlled release of amphotericin b for dermal local antifungal therapy of candidiasis. *Eur. J. Pharm. Sci.* **2020**, *145*, 105255. [CrossRef]
156. De Siqueira, E.C.; de Souza Rebouças, J.; Pinheiro, I.O.; Formiga, F.R. Levan-based nanostructured systems: An overview. *Int. J. Pharm.* **2020**, *580*, 119242. [CrossRef]
157. Erginer, M.; Akcay, A.; Coskunkan, B.; Morova, T.; Rende, D.; Bucak, S.; Baysal, N.; Ozisik, R.; Eroglu, M.S.; Agirbasli, M. Sulfated levan from halomonas smyrnensis as a bioactive, heparin-mimetic glycan for cardiac tissue engineering applications. *Carbohydr. Polym.* **2016**, *149*, 289–296. [CrossRef]
158. Gomes, T.D.; Caridade, S.G.; Sousa, M.P.; Azevedo, S.; Kandur, M.Y.; Öner, E.T.; Alves, N.M.; Mano, J.F. Adhesive free-standing multilayer films containing sulfated levan for biomedical applications. *Acta Biomater.* **2018**, *69*, 183–195. [CrossRef]
159. Francesko, A.; Tzanov, T. Chitin, chitosan and derivatives for wound healing and tissue engineering. In *Biofunctionalization of Polymers and their Applications*; Springer: Berlin, Germany, 2010; pp. 1–27.
160. Elieh-Ali-Komi, D.; Hamblin, M.R. Chitin and chitosan: Production and application of versatile biomedical nanomaterials. *Int. J. Adv. Res.* **2016**, *4*, 411.
161. Poth, N.; Seiffart, V.; Gross, G.; Menzel, H.; Dempwolf, W. Biodegradable chitosan nanoparticle coatings on titanium for the delivery of bmp-2. *Biomolecules* **2015**, *5*, 3–19. [CrossRef] [PubMed]
162. Mishra, S.K.; Ferreira, J.; Kannan, S. Mechanically stable antimicrobial chitosan–pva–silver nanocomposite coatings deposited on titanium implants. *Carbohydr. Polym.* **2015**, *121*, 37–48. [CrossRef]

163. Bernkop-Schnürch, A.; Dünnhaupt, S. Chitosan-based drug delivery systems. *Eur. J. Pharm. Biopharm.* **2012**, *81*, 463–469. [CrossRef] [PubMed]
164. Li, J.; Wang, Y.; Liang, R.; An, X.; Wang, K.; Shen, G.; Tu, Y.; Zhu, J.; Tao, J. Recent advances in targeted nanoparticles drug delivery to melanoma. *Nanomed. Nanotechnol. Biol. Med.* **2015**, *11*, 769–794. [CrossRef] [PubMed]
165. Guadagnini, R.; Halamoda Kenzaoui, B.; Walker, L.; Pojana, G.; Magdolenova, Z.; Bilanicova, D.; Saunders, M.; Juillerat-Jeanneret, L.; Marcomini, A.; Huk, A. Toxicity screenings of nanomaterials: Challenges due to interference with assay processes and components of classic in vitro tests. *Nanotoxicology* **2015**, *9*, 13–24. [CrossRef] [PubMed]
166. Davies, H.; Bignell, G.R.; Cox, C.; Stephens, P.; Edkins, S.; Clegg, S.; Teague, J.; Woffendin, H.; Garnett, M.J.; Bottomley, W. Mutations of the braf gene in human cancer. *Nature* **2002**, *417*, 949–954. [CrossRef]
167. Huang, T.; Karsy, M.; Zhuge, J.; Zhong, M.; Liu, D. B-raf and the inhibitors: From bench to bedside. *J. Hematol. Oncol.* **2013**, *6*, 30. [CrossRef]
168. Maertens, O.; Kuzmickas, R.; Manchester, H.E.; Emerson, C.E.; Gavin, A.G.; Guild, C.J.; Wong, T.C.; De Raedt, T.; Bowman-Colin, C.; Hatchi, E. Mapk pathway suppression unmasks latent DNA repair defects and confers a chemical synthetic vulnerability in braf-, nras-, and nf1-mutant melanomas. *Cancer Discov.* **2019**, *9*, 526–545. [CrossRef]

Review

The Pulsed Electron Deposition Technique for Biomedical Applications: A Review

Anna Liguori [1], Chiara Gualandi [1,2], Maria Letizia Focarete [1,3], Fabio Biscarini [4,5] and Michele Bianchi [4,*]

1 Department of Chemistry "Giacomo Ciamician" and INSTM UdR of Bologna, University of Bologna, via Selmi 2, 40126 Bologna, Italy; anna.liguori@unibo.it (A.L.); c.gualandi@unibo.it (C.G.); marialetizia.focarete@unibo.it (M.L.F.)
2 Interdepartmental Center for Industrial Research on Advanced Applications in Mechanical Engineering and Materials Technology, CIRI-MAM, University of Bologna, Viale Risorgimento 2, 40136 Bologna, Italy
3 Health Sciences and Technologies (HST) CIRI, University of Bologna, Via Tolara di Sopra 41/E, Ozzano dell'Emilia, 40064 Bologna, Italy
4 Center for Translational Neurophysiology of Speech and Communication, Istituto Italiano di Tecnologia, via Fossato di Mortara 17–19, 44121 Ferrara, Italy; fabio.biscarini@iit.it
5 Life Science Department, University of Modena and Reggio Emilia, via Campi 103, 41125 Modena, Italy
* Correspondence: michele.bianchi@iit.it

Received: 11 December 2019; Accepted: 23 December 2019; Published: 25 December 2019

Abstract: The "pulsed electron deposition" (PED) technique, in which a solid target material is ablated by a fast, high-energy electron beam, was initially developed two decades ago for the deposition of thin films of metal oxides for photovoltaics, spintronics, memories, and superconductivity, and dielectric polymer layers. Recently, PED has been proposed for use in the biomedical field for the fabrication of hard and soft coatings. The first biomedical application was the deposition of low wear zirconium oxide coatings on the bearing components in total joint replacement. Since then, several works have reported the manufacturing and characterization of coatings of hydroxyapatite, calcium phosphate substituted (CaP), biogenic CaP, bioglass, and antibacterial coatings on both hard (metallic or ceramic) and soft (plastic or elastomeric) substrates. Due to the growing interest in PED, the current maturity of the technology and the low cost compared to other commonly used physical vapor deposition techniques, the purpose of this work was to review the principles of operation, the main applications, and the future perspectives of PED technology in medicine.

Keywords: pulsed electron deposition; thin films; orthopedic applications; bioactivity; ceramic coatings; yttria-stabilized zirconia; calcium phosphates; hydroxyapatite; biomimetic coatings; antibacterial coatings

---

1. Introduction

In the frame of physical vapor deposition (PVD) techniques, pulsed electron deposition (PED) is a well-established technology to fabricate thin films for photovoltaic, superconductor, and optoelectronic applications [1–7]. PED technology belongs to the family of the channel spark discharges, a type of hollow cathode glow discharge in which a target material is ablated by the local heating induced by an accelerated electron beam [8,9]. At the very beginning, this technique was mainly employed for the deposition of both inorganic, i.e., superconductive $MBa_2Cu_3O_{7-x}$ [10], and organic, i.e., polytetrafluoroethylene (PTFE) [11], thin films.

The modifications introduced by Taliani and coworkers, aimed both at improving the reliability of the technique and at increasing the electron beam density [12,13], identify the transition of the PED technology from the traditional channel spark discharge to the pulsed plasma deposition technique

(PPD). Similar to the earlier technology, PPD bases on the extraction of electrons from a plasma generated in a rarefied gas through a narrow channel dielectric [14] (Figure 1). The electron beam generation system mainly comprises a hollow cathode connected to the electrical generator, a ring acting as the anode, and a dielectric tube connecting the cathode and the anode and extending beyond the anode [14]. The negative charges of the generated plasma are accelerated, through the application of an up to 20 kV potential difference between the hollow cathode and the anode, and channeled into the dielectric tube in order to reach the target placed at the right opposite end of the tube [14]. The target material impacted by the fast (200–400 ns) pulsed electrons is emitted in the form of a highly ionized plasma, called plasma plume, with the axis perpendicular to the surface of the target. The plasma plume enters in contact with the substrate where the material is, therefore, deposited, forming a film [14].

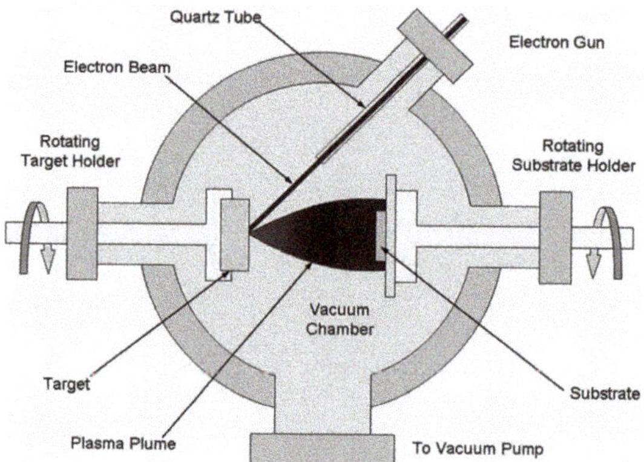

Figure 1. Scheme of the pulsed plasma deposition process, reproduced from [15] with permission from the Royal Society of Chemistry.

Conceptually similar to the widely exploited pulsed laser deposition (PLD), PPD shows additional advantages, mainly due to the replacement of the short pulse of photons in PLD with a short pulse of electrons. Indeed, with respect to PLD, PPD also enables one to process transparent or highly reflective materials and it exhibits higher electrical efficiency (30%) and lower capital costs [16]. A further advantage of PPD is the possibility to produce films with a controlled stoichiometry of the target, even for complex compounds, such as $La_{0.7}Sr_{0.3}MnO_3$ [2]. In the framework of photovoltaic and optoelectronic applications, PPD has been employed for the deposition of a wide range of inorganic thin films, such as tungsten-doped thin oxide [1,4], pyrite [17], and transparent conductive p-type lithium doped [6] and potassium-doped [7] nickel oxide. Efforts have been also devoted to investigating the role of PPD operating conditions (i.e., substrate temperature, electron beam energy) and of the target's characteristics (such as density and composition) on the properties of the deposited coatings [18–20]. Finally, studies have aimed to evaluate the suitability of PPD technology for industrial applications, and concluded that the contextual use of multiple targets together with the active cooling of the electrodes and the target lead to a reduction of the deposition time and to an improvement of the uniformity of the coating's thickness [18,21].

More recently, a PED evolution named ionized jet deposition (IJD), entered the thin film technology market, with the explicit promise of overtaking the main drawbacks of PPD, such as the need for periodically substituting the dielectric tube confining the electron beam till the target surface, and providing, at the same time, superior power and robustness [22]. The IJD working principle is sketched in Figure 2: IJD is based on a pulsed electron source able to generate in vacuum ultra-short electric discharges in the megawatt range (MW) range, produced with a high voltage pulse amplitude up to 25 kV and a duration lower than 1 μs. The discharge, supported by a gas jet, is directed into a solid target through a system of trigger and auxiliary electrodes, generating a superficial explosion with the consequent emission of material in the form of a plasma. The plasma emitted by the target produces on the substrate, a dense coating presenting the same composition of the target [22].

With respect to PPD, IJD is characterized by a higher efficiency (82%–88%) [22] and leads to a more optimized stoichiometry conservation, turning out to be particularly suitable for: (i) the deposition of superconducting oxides or photoactive semiconductors, (ii) the production of crystalline thin films, and (iii) deposition on any kind of substrate, including thermo-sensitive materials. IJD has been tested for the deposition of SnS thin film layers, useful for photovoltaic applications [23,24]; amorphous chameleon coatings, namely, hard metal carbides and nitrides [25]; and macroscopically homogeneous, adhesive, and cross-linked poly(methylmetacrylate), polystyrene, polyvinylchloride coatings on stainless steel, and glass substrates [26].

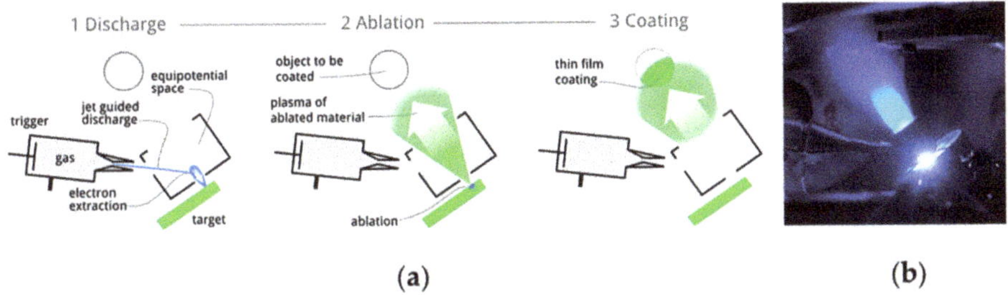

Figure 2. (**a**) Scheme of the ionized jet deposition process; (**b**) system in operation [22].

As briefly summarized, since its introduction, PED has been exclusively exploited to obtain inorganic and organic coatings for photovoltaic or optoelectronic applications. No biomedicine-related applications of this technology were proposed until 2013 [14]. It should be noted that other, more established techniques have been used, since it takes many years to modify the surface of metals and metal alloys by making use of an electron beam [27]; however, due to the very different working principle compared to PED, these techniques will not be discussed within this review.

2. Scope and Methodology of the Review

Moved by the continuous demands for innovative yet sustainable approaches to improve the mechanical and biological performances of common implantable materials in orthopedic and dental fields, a number of PVD technologies have been tested over the years to endow the implants with functional coatings, including PLD [28], magnetron sputtering [29], electron beam physical vapor deposition [30], arc ion plating [31], and cathodic arc deposition [32].

In this bio-context, PED has been only recently explored, triggered by the opportunity to obtain well-adhered, nanostructured, and stoichiometric thin films, and coatings on metallic, ceramic, or plastic substrates, along with lower costs for industrial up-scaling compared to competitor techniques. The aim of this work was, thus, to review and discuss the main scientific literature concerning bio-applications of the PED technology, in order to highlight the potentialities and main drawbacks, and point out future directions for the spread of this breakthrough technology in different biomedical sectors.

A systematic search using Google Scholar, Scopus, Web of Science, and Google Patents from 1996 until November 2019 was performed. Principal search strings to filter among the different techniques were: "channel spark discharge", "channel spark ablation", "pulsed plasma deposition", "pulsed electron deposition", "pulsed electron ablation", and "ionized jet deposition". Each of these terms was searched together with one or more of the following terms: "biomedical applications", "ceramic", "(hard) coatings (or thin films)", "load bearing", "articulating surfaces", "orthopedic (or dental) application", "bioactive", "hydroxyapatite", "calcium phosphate", "biogenic", "antibacterial", "antimicrobial", and "antiviral". Finally, to find additional papers possibly missed through the database searches, the list of citations from each paper was reviewed.

3. Hard Ceramic Coatings for Articulating Surfaces in Joint Arthroplasty

The first reported biomedical application of PED technology is placed in the context of orthopedic prostheses for total joint arthroplasty (TJA) [15]. Nowadays, the vast majority of orthopedic prostheses consist of an ultra-high molecular weight polyethylene (UHMWPE) inlay (or its cross-linked variant, XPE) [33] articulating against a cobalt-chromium or titanium alloy component. Due to osteolysis and aseptic loosening of an implant [34], which is highly correlated to the formation of wear debris associated to the UHMWPE insert [35–37], the average lifetime for a TJA implant is of about 10 years [38]; then, a revision surgery is often required.

PED was formerly used to deposit, at room temperature, crystalline yttria-stabilized zirconia (YSZ) coatings directly on the surface of the UHMWPE inlay, with the aim of preserving it from wear and plastic deformation; thus, increasing the lifetime of the implant overall [15]. In a series of explorative works, micrometric-thick YSZ coatings showing a nanostructured surface texture (Figure 3) were deposited using a PPD setup [15]. YSZ coatings deposited at room temperature exhibited a cubic crystalline phase, conferring high fracture toughness and excellent adhesion to the plastic substrate, despite the significant mechanical mismatch between the harder coating and the softer substrate [15,39]. In addition, nanoindentation tests indicated a much higher strength of normal plastic deformation for YSZ-coated UHMWPE compared to uncoated UHMWPE, along with a drastically reduced creep of the plastic inlay [40]. In the same work, the authors investigated more into detail, the effect of the working gas pressure on the final characteristics of the coatings. Interestingly, they found that the YSZ coatings deposited at a lower working gas pressure (i.e., 5.5×10^{-3} mbar) were less subjected to plastic deformation than those deposited at a higher pressure (i.e., 8.0×10^{-3} mbar) [40], due to a higher indentation hardness of the latter.

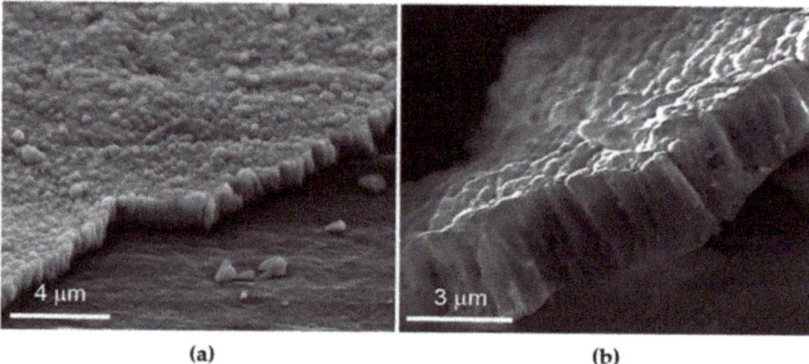

Figure 3. Scanning electron microscope (SEM) images of yttria-stabilized zirconia (YSZ) coatings deposited on ultra-high molecular weight polyethylene (UHMWPE): (**a**) microphotograph of a ≈1.7 µm thick coating; (**b**) detail of the cross-section of a ≈3 µm thick coating, reproduced from [15] with permission from the Royal Society of Chemistry.

In a subsequent study, finite element analysis (FEA) was used to determine the optimal coating's thickness in order to minimize UHMWPE strain, crack onset, and stresses at the coating-substrate interface [41]. Both nanoindentation, to simulate the application of high local loads, and micro-indentation tests, to investigate the adhesion of brittle coatings on ductile substrates, were simulated by FEA. The simulations indicated that the local stresses required to promote coating failure strictly depends on its thickness. Specifically, the authors found that thinner coatings (≈0.5 µm) were more compliant to the applied load compared thicker coatings (2 ÷ 4 µm), as an increase of the thickness led to a decrease of the contact area between the coating and the substrate and an increase of the corresponding minimum principal stress (i). The value of the maximum principal stress was instead achieved at the substrate-coating interface; therefore, a stress applied in this region may lead to coating delamination (ii) [41].

PPD has also been tested for depositing YSZ coatings on metal substrates, such as AISI 316-L, titanium, and titanium alloys, to indirectly protect the UHMWPE inlay from wear by coupling it with a low-friction counterpart [42–45]. In a series of papers, where a range of deposition parameters was investigated, the authors reported the possibility to reduce the wear rate of UHMWPE by finely tuning of the working gas pressure [42]. Indeed, gas pressures in the range (4–8) $\times 10^{-3}$ mbar provided low roughness and high thickness coatings on AISI 316-L, capable of ensuring a reduced wear rate and a steady friction coefficient during the tribological test, both under dry and lubricated testing conditions [42]. Along with these results, the deposition of YSZ onto titanium and titanium alloy substrates, carried out with a working pressure of 6×10^{-3} mbar, significantly reduced the UHMWPE wear rate [43–45] (17% and 4% under dry and lubricated conditions, respectively) of the coated metallic substrate with respect to the uncoated one [43]. This behavior was ascribed to the reduction of the contact area between the polymeric component and the coated metal compared to the uncoated one, strongly limiting the formation of UHMWPE debris, which is related to severe abrasive and adhesive wear at the titanium counter-face [43,46,47].

In a subsequent study, the viabilities of mesenchymal stem cells (MSCs) and pre-osteoblast MC3T3-E1 cells on coated and uncoated titanium substrates [45] were investigated. Cell proliferation increased from day 1 to day 7 for both the samples and the cell types (Figure 4). On day 7, cell proliferation was higher on coated samples than on uncoated ones; further, no changes in the morphology of the nuclei nor abnormal alterations in MSCs and MC3T3 cells seeded on coated samples were detected [45], suggesting that the nanostructured YSZ coatings deposited by PED were non-cytotoxic.

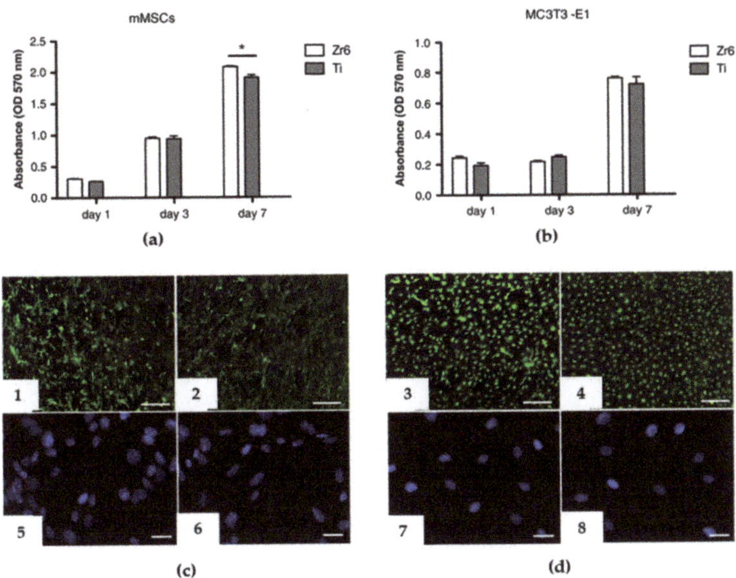

Figure 4. Analysis of mesenchymal stem cell (MSC) (**a**) and MC3T3-E1 (**b**) cell proliferations by MTT assay after 1, 3, and 7 days of seeding on samples; * $p \leq 0.01$; $n = 3$. Analysis of cell viability was carried out through live/dead assay (**c**,**d**); calcein AM stains for live cells in green; EthD-1 stains for dead cells in red: (1) and (2) show MSCs; (3) and (4) show MC3T3-E1 seeded on coated and uncoated samples, respectively, after 3 days. Nuclear shape of MSCs (5,6) and MC3T3-E1 (7,8) on coated (5,7) and uncoated (6,8) samples. Zr6 = coated samples; Ti = uncoated samples. Scale bars: 200 μm (1–4), 25 μm (5–8) [45]. Reprinted with permission from [45]. © 2016 Springer Nature.

In a very recent study, nanostructured YSZ thin films have been deposited by IJD onto different materials (i.e., glass, silicon, titanium, and poly(etheretherketone) (PEEK)), with the aim of investigating the effect of the roughness of the substrate on the microstructure of the growing film. The authors found that, while "smooth" substrates (glass and silicon), did not affect the film roughness, "rough" surfaces (titanium and PEEK) led to homogeneous films well resembling the topography of the substrate [48].

Zirconia-toughened-alumina (ZTA) is a ceramic composite material particularly appealing for dental and orthopedic applications due to its high strength, fracture toughness, elasticity, hardness, and wear resistance [49,50]. ZTA thin films were deposited by PPD onto the surface of AISI 420 substrates and showed high adhesion to the substrate, as documented by the relatively high load (0.8 N) required to achieve the crack onset in the coating during the micro-scratch testing, and the low friction coefficient when tested under lubricated conditions (0.12–0.15) [51]. Further on, X-ray diffraction (XRD) analysis indicated that ZTA coatings exhibited a mixed phase of tetragonal zirconia and α-phase alumina embedded in a matrix of amorphous alumina [52].

4. Bioactive Coatings and Thin Films by PED

During the last few decades, great effort has been dedicated to improving the integration of orthopedic and dental implants with the surrounding bone tissue (a process named osseointegration) [53–55]. One of the most investigated strategies relies on the coating of the implant surface with materials capable of triggering the biological cascade of events eventually leading to osseointegration (i.e., bioactive materials) [56]. Due to the chemical similarity with the inorganic phase of bone, calcium phosphates (CaPs) and hydroxyapatite (HA) in particular, have been extensively investigated both in vitro and in animal models [57].

In this context, PED technology has been tested for the deposition of nanostructured bioactive thin films on metallic and polymeric implants. In a first study, CaP films were deposited on titanium alloy substrate by ablating a sintered HA target; the analysis of morphology, microstructure and in vitro cytotoxicity were carried out focusing in particular on the effect of the post deposition thermal treatment performed at 600 °C for 1 h [58]. The treatment led to crystalline HA and improved the mechanical properties of the film compared to the nearly amorphous as-deposited one, for instance leading to increased elastic strain to failure, resistance to plastic deformation, and adhesion to the substrate. Besides, no significant differences were found in terms of cytotoxicity between as-deposited and treated films, both being capable of subtly promoting adhesion and proliferation of primary osteoblast cells [58].

Among the different implant materials, poly(etheretherketone) (PEEK) is increasingly used as an alternative to metals, especially for spinal fusion, due to excellent resistance to chemicals and sterilization process, and mechanical properties [59–62]. Nevertheless, PEEK totally lacks bioactivity (bioinert material) [63,64]. IJD has been used to deposit strontium doped-hydroxyapatite films on the PEEK substrates with the aim of improving its bioactivity [65]. Strontium ion (Sr^{2+}) has been selected as dopant of HA because of its well-known osteogenic effect; i.e., its capability to promote osteoblasts activity (new bone apposition) and inhibit osteoclasts activity (old bone resorption) [66]. Similar to the previously reported study, a two-step approach was adopted. Different Sr/Ca molar ratio targets (from 0.04 to 0.20) were fabricated and ablated to obtain Sr-HA films with different content of Sr. Energy-dispersive X-ray spectroscopy (EDX) microanalysis confirmed the high degree of fidelity of stoichiometry transfer from the target material to the film. Films were characterized by dense and uniform packing of spherical-like micrometric and submicrometric grains, the size of which ranged mainly from 50 to 100 nm (Figure 5a,b) [65]. The formation of this nanostructures should be emphasized, since nanotopography plays an important role on promotion of new bone apposition [67]. Noteworthy is that the SrCaP coatings increased the surface roughness (Figure 5c) while decreasing the water contact angle (Figure 5d), corresponding to a higher wettability compared to hydrophobic PEEK. This should be remarked on, as the improvement of PEEK surface roughness and wettability is envisioned to promote adhesion, spreading, proliferation, and differentiation of bone cells [68].

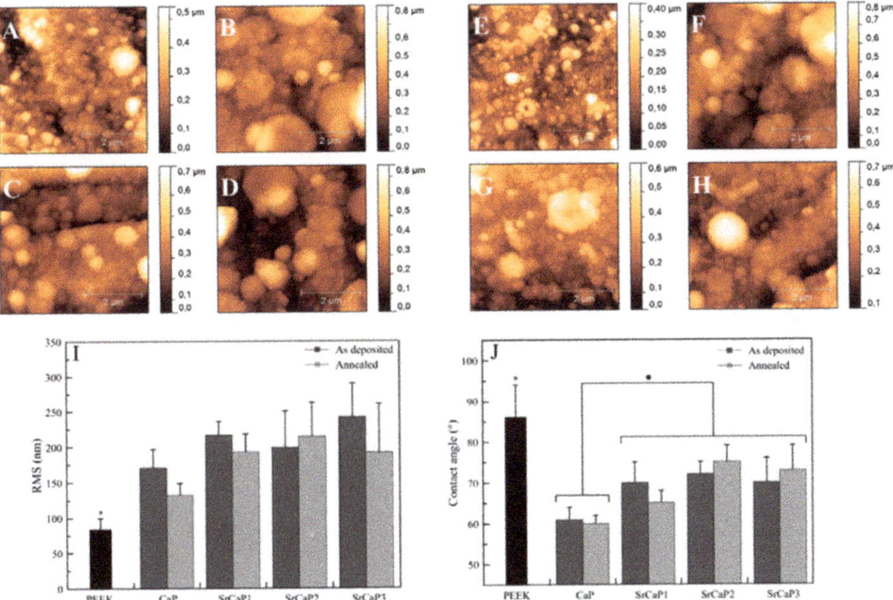

Figure 5. Atomic force microscope (AFM) images of the as-deposited (**A–D**) and annealed (**E–H**) coatings; CaP (Sr^{2+}-free coating) (**A**,**E**), SrCaP1 (Sr/Ca molar ratio = 0.04) (**B**,**F**), SrCaP2 (Sr/Ca molar ratio = 0.10) (**C**,**G**), and SrCaP3 (Sr/Ca molar ratio = 0.20) (**D**,**H**). Root Mean Square (RMS) roughness (**I**) and water contact angle (**J**) of the bare PEEK, as-deposited, and annealed coatings. Reprinted with permission from [65]. © 2017 Elsevier.

In the field of bioactive coatings, biomimetic coatings resembling the composition of natural bone apatite, which is indeed a calcium-deficient, highly-substituted, and poorly crystalline hydroxyapatite [69], are strongly desired, as they are expected to better promote differentiation implant osseointegration compared to highly-crystalline stoichiometric HA or other CaP phases [70–72]. Aiming to fabricate highly biomimetic coatings, bone apatite-like (BAL) thin films were deposited on titanium substrate for the first time by direct ablation of a biogenic source; i.e., a deproteinized bovine bone shaft, by making use of IJD technology [73,74]. EDX indicated that the composition of BAL thin films well mimicked that of the biogenic target, particularly with regard to Ca/P ratio and the amount of Mg and Na ions. Remarkably, treatment of the as-deposited highly amorphous films at 400 °C for 1 h promoted an increase in crystallinity [73,74] (Figure 6). Indeed, XRD patterns of as-deposited films showed nanocrystalline domains (HA, Figure 6a), or an essentially amorphous phase (BAL, Figure 6b) consistent with the different crystalline degree of the targets. After annealing, films crystallized in a phase close to the natural apatite structure [75].

Demonstrations of high adhesion of the BAL films onto the surfaces of 3D objects, such as dental screws, have been also reported [74].

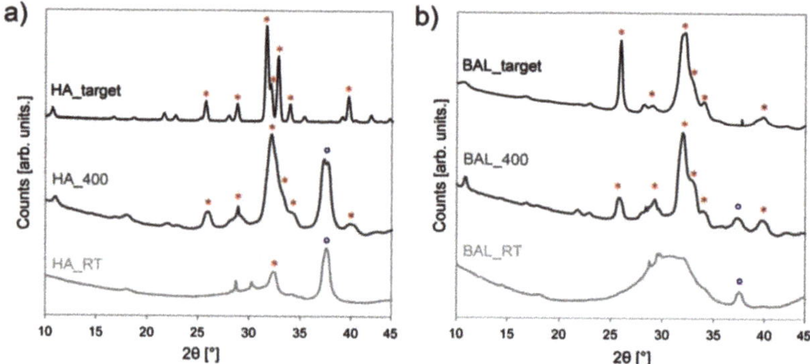

Figure 6. XRD spectra of the hydroxyapatite (HA) (**a**) and bone apatite-like (BAL) (**b**) targets; as-deposited and annealed films. Reprinted with permission from [73]. © 2017 Elsevier.

In light of these results, the adhesion, proliferation, and osteogenic differentiation of human dental pulp stem cells (hDPCs) on BAL thin films were evaluated and compared with those achieved on stoichiometric HA thin films, either as-deposited or after the thermal treatment [76]. Results well demonstrated that hDPSCs cultured on annealed BAL thin films preserved their morphology and homogeneously proliferated through the whole surface when compared with the other films, as shown in Figure 7a,c,e,g [76]. Furthermore, cells grown in an osteogenic medium on annealed BAL thin films were able to express osteogenic markers, reaching a higher commitment towards osteogenic differentiation with respect to as-deposited and annealed HA and as-deposited BAL films [76].

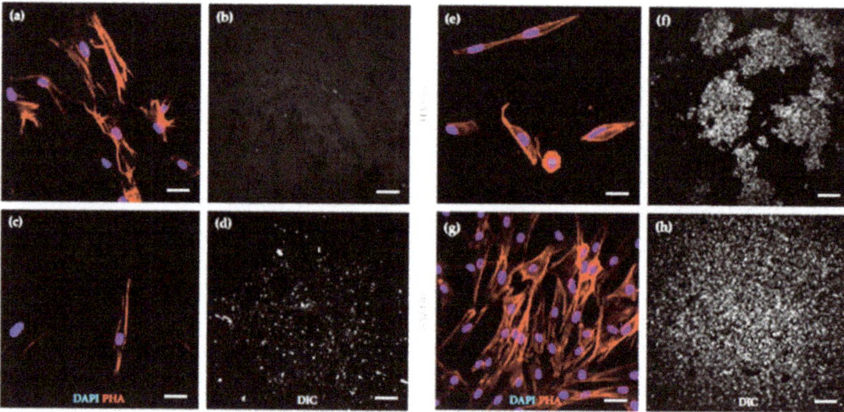

Figure 7. Immunofluorescence images (**a**,**c**,**e**,**g**) showing cell morphology of hDPSCs cultured on HA and BAL films and differential interface contrast microscopy images (**b**,**d**,**f**,**h**) representing film adhesion to the underlying surface at 7 days of culturing. Immunofluorescence analysis was performed against phalloidin (PHA). Nuclei were counterstained with 4′,6-diamidino-2-phenylindole (DAPI). Scale bar is 10 μm [76].

In the framework of biogenic coatings, PED has been used also for the fabrication of thin films made of bioglass [77]. Thin films of 45S5 and CaO-rich (named CaK) bioglasses were deposited by IJD on titanium alloy substrates [78]. The effect of the thermal annealing performed for 1 h at 600 °C and 750 °C for 45S5 and CaK, respectively, was also investigated [78]. The surface morphology of the as-deposited and annealed coatings was characterized by the presence of spherical aggregates, while the chemical composition of the films was very close to that of the bioactive glass target (Figure 8) [78].

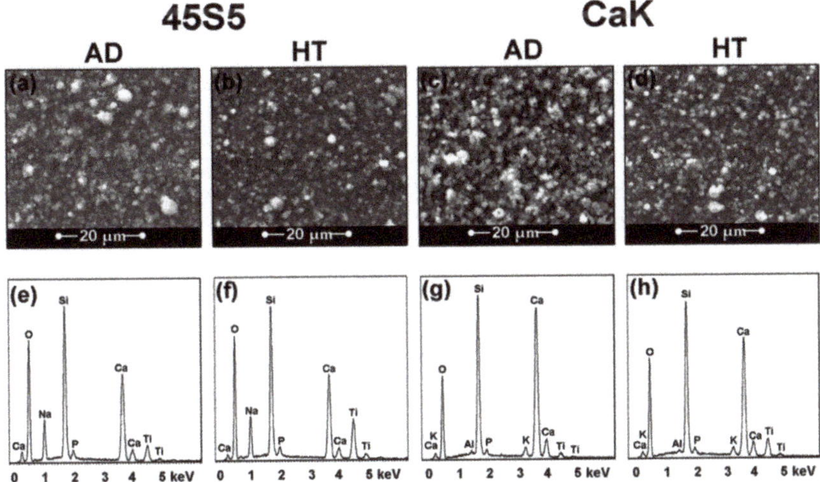

Figure 8. SEM images of as-deposited (**a**,**c**) and thermal annealed (**b**,**d**) coatings; results of EDX analysis on as-deposited (**e**,**g**) and thermal annealed (**f**,**h**) coatings. Reprinted with permission from [78]. © 2017 Elsevier.

Notably, the resistances to delamination of the bioglass films were found to closely correlate to their crystallinity. The 45S5 annealed coatings were poorly crystalline, and consequently, exhibited poor resistance, as can be observed by evaluating the critical loads of partial and complete delamination and the worn track from the micro-scratch test (Figure 9). On the contrary, CaK annealed coatings were more crystalline and showed higher resistance to delamination [78].

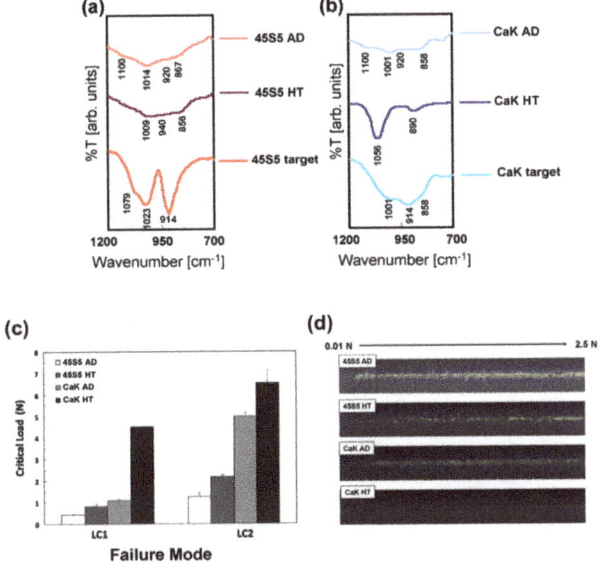

Figure 9. Fourier-transform infrared spectroscopy (FT-IR) spectra of 45S5 (**a**) and CaK (**b**) targets and for as-deposited and heat-treated coatings. Critical loads for partial (LC1) and complete (LC2) delamination (**c**) and optical images (objective magnitude: 5×) of the worn track (**d**) obtained by the micro-scratch test.

5. Antibacterial Thin Films by PED

About four million healthcare-related infections occur per year in Europe, causing about 37,000 deaths, 16 million additional days of hospitalization, and a direct cost of seven billion Euros [79]. Infections can start from microbial adhesion on the implant surface immediately upon insertion, during surgery, because of hematogenous spreading of the microorganisms to the implant, or because of spreading from an adjacent infection site, leading, in any case, to the formation of a biofilm. Upon formation of the biofilm, microorganisms are well protected from the host immune system and from antibiotics, so they become capable of proliferating and invading other implants or adjacent bone sites [80]. If biofilm formation occurs, eradication of the infection requires implant removal and revision surgery, and, in some cases, can have dramatic consequences for the patient. Currently, the most successful strategy for preventing infections is systemic antibiotic therapy, which, however, increases the risk of systemic toxicity and of the development of resistant bacterial strains [81]. For this reason, inorganic antimicrobial coatings able to locally deliver antibacterial agents directly in the site of insertion without causing bacterial resistance, are extremely appealing [82,83]. In addition, antimicrobial coatings must not only be effective and biocompatible, but must also not impair osseointegration; i.e., the direct structural and functional contact between the implant and the host bone [84].

A possible strategy is to coat implantable materials with materials eliciting both bioactive and antibacterial effects through the incorporation of inorganic factors such metal nanoparticles in bioactive coatings [85–87]. Silver (Ag) is undoubtedly the workhorse among the many inorganic antibacterial species, thanks to its widely documented wide spectrum anti-microbial and anti-viral efficacy [88,89]. Recently, efforts have been focused to the fabrication of nanostructured Ag thin films to exploit the enhanced antibacterial properties of nanostructured materials compared to micro-structured materials [90–92].

The suitability of IJD technology for the fabrication of Ag nanostructured thin films onto titanium alloy and silicon oxide substrates was explored [93]. By ablating a pure metallic Ag target, IJD provided pure metallic Ag films which were nanostructured and highly homogeneous (Figure 10). Notably, differently from ceramics, the high thermal conductivity and density of the Ag target (about three times higher than that of BAL and YSZ) allow one to deposit very smooth films, with nanometric corrugation (mean surface roughness in the range 0.7–1.2 nm), with low amounts of particulate and a mean size of the grains of few nanometers [93]. Further, AFM highlighted that no significant variation of the grain size could be observed by increasing the film thickness (Figure 10).

For biomedical applications in which the control over the morphology is fundamental for the regulation of the balance between cytotoxicity and antimicrobial properties, the controlled dewetting of Ag thin films allowed some to obtain non-cytotoxic structures presenting antimicrobial properties [94,95]. The dewetting of thin films is a thermally-driven process, which, performed at a temperature below the melting point, leads to the evolvement of the surface morphology up to the formation of separate droplets/3D islands [96–98]. The dewetting process was carried out in air for 1 h with temperature varying from 200 to 600 °C on Ag thin films deposited by means of IJD technology onto silica substrate [99]. The morphological analysis highlighted that for dewetting process performed at 200 °C, the surface of the film was still continuous and only a few hillocks were found. Further, temperature increased up to 300 °C promoted the formation of micrometric holes and the partial exposure of the silica substrate; by further heating the film up to 600 °C the formation and separation of 3D islands was detected [99].

Among inorganic antibacterial materials, magnetite nanoparticles were demonstrated to impair biofilm growth by rapidly penetrating the biofilm pores in the presence of an external magnetic field [100]. HA/magnetite coatings have been fabricated by PPD. The composition well resembled the one of the target, although a random and inhomogeneous distribution of the magnetic particles on the film's surface was highlighted by both electrostatic force microscopy and scanning tunneling microscopy [101]. The antibacterial tests, performed by evaluating the adhesion of *Escherichia coli* after 4 h of incubation on HA/magnetite and HA coatings with and without the presence of an external magnetic field (intensity 1.45 T), demonstrated the suitability of the HA/magnetite coatings in hindering the bacterial adhesion in comparison to HA coatings, independently from the application of the magnetic field [101]. According to the previous results [100], the presence of the magnetic field, induced by a permanent magnet placed below the samples, enhanced the anti-adhesive properties of the HA/magnetite coatings [101].

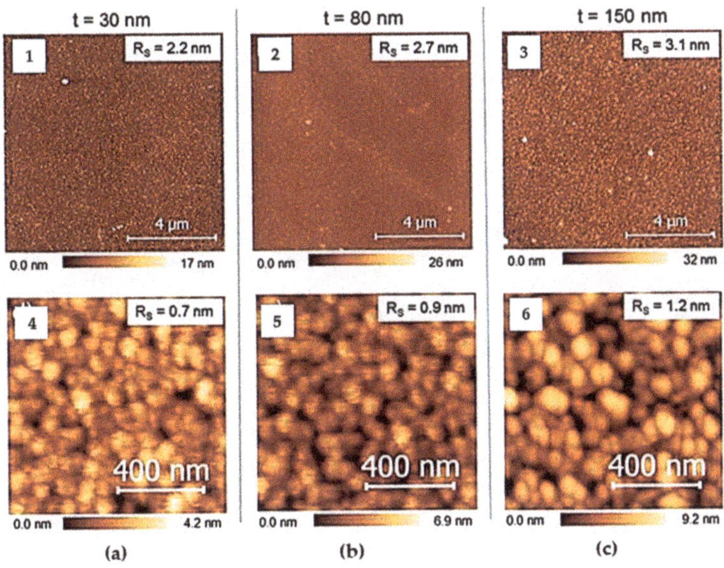

Figure 10. AFM images (10 µm × 10 µm) of Ag coatings deposited on Si at three different thickness values: (**a**) t = 48 ± 8, (**b**) 78 ± 12, and (**c**) 159 ± 19 nm. RMS roughness (R_s) values are indicated in the label for each image. Reprinted with permission from [93]. © 2019 Elsevier.

6. Summary and Future Perspectives

Despite the fact that PED has been used for several decades, application of PED to depositing coatings and thin films of biomedical interest is relatively new. However, in less than a decade, a significant number of manuscripts concerning the use of PED in the biomedical field have been published (Table 1), likely due to the entry in the market of more and more robust versions of the technology, such as IJD.

Table 1. Summary of the available literature on PED technology for biomedical applications.

Hard Coatings for Articulating Surfaces			
Target	**Substrate**	**Coating's Main Features**	**References**
ZrO_2 stabilized with Y_2O_3	UHMWPE	Fully cubic crystalline phase; excellent adhesion to the plastic substrate; high strength to normal plastic deformation.	[31,39,40]
ZrO_2 stabilized with Y_2O_3	AISI 316-L	Low roughness and high thickness; reduction of the wear rate of polymeric counterpart; steady friction coefficient during the tribological test.	[42]

Table 1. *Cont.*

Hard Coatings for Articulating Surfaces			
Target	**Substrate**	**Coating's Main Features**	**References**
ZrO_2 stabilized with Y_2O_3	Titanium and titanium alloys	Significant reduction of UHMWPE wear rate; prevention of the formation of abrasion scratches on the substrate; no cytotoxic effects.	[43–45]
ZrO_2 stabilized with Y_2O_3	Glass, silicon, titanium, PEEK	Morphology independent from the substrate in case of flat substrates; morphology resembling the one of the substrate in case of rough substrates.	[48]
75% Al_2O_3/25% ZrO_2	AISI 420	High adhesion to the substrate; increase of the surface roughness along with the thickness.	[51,52]
Bioactive Coatings and Thin Films			
Target	**Substrate**	**Coating's Main Features**	**References**
Sintered HA	Titanium alloy	Transition from amorphous to crystalline coatings by means of the thermal treatment of the deposited coating; cell adhesion and proliferation on both as-deposited and annealed coatings.	[58]
CaP target doped with Sr^{2+}	PEEK	Same stoichiometry of the target; dense and uniform packing of spherical-like grains; Sr^{2+} in the coating increased the surface roughness and decreased the water contact angle.	[65]
Biogenic HA (deproteinized bovine bone shaft)	Titanium	Composition well mimicking that of the target; increase in crystallinity up to that typical of biogenic apatite through the thermal treatment of the as deposited films; high adhesion on the surface of 3D objects.	[73,74]
Biogenic HA (deproteinized bovine bone shaft)	Glass	hDPSCs cultured on annealed films preserved their morphology and homogeneously proliferated through the whole surface; cells grown in an osteogenic medium on annealed thin films expressed osteogenic markers.	[76]
45S5 and CaK bioglasses	Titanium alloy	Spherical aggregates on both as-deposited and annealed coatings; chemical composition analogous to that of the starting bioactive glasses; resistance to delamination closely correlated to the crystallinity: the higher the crystallinity of the coating the higher the resistance.	[78]

Table 1. *Cont.*

Antibacterial Thin Films			
Target	**Substrate**	**Coating's Main Features**	**References**
Ag	Titanium alloy and silicon dioxide	Composition resembling the one of the target; homogeneity in thickness; presence of grains composed by small subunits with a size of few nm; no variation of the grains' diameter with the increase of the film thickness.	[93]
Ag	Silicon dioxide	Formation and separation of 3D islands from the coating when subjected to a dewetting process carried out at 600 °C for 1 h.	[99]
HA/magnetite	Silicon wafers	Composition resembling the one of the target, despite the random and inhomogeneous distribution of the magnetic particles on the surface; the coatings hindered the *E. coli* adhesion.	[101]

The deposition of functional metallic coatings represents one of the most promising research fields in which PED technology can provide highly performing solutions. For instance, due to its suitability to process thermosensitive substrates, PED can be used to deposit antibacterial coatings onto electrospun polymeric substrates with the aim to produce patches for wound healing and tissue regeneration (unpublished results by the authors). Furthermore, PED technology can be exploited to coat polymeric fibers with conductive materials for wearable electronics [102] and electrically conductive textiles [103], where flexible and well-adhesive conductive coatings are desired.

Besides, the deposition of bioactive or fully biomimetic coatings and the promising results obtained from the in vitro tests in terms of bioactivity, mechanical properties, and adhesion, certainly represent a valuable starting point for further in vivo investigations, aside from a novel promising strategy for endowing bio-inert implantable scaffolds with novel functional properties [104,105].

From that perspective, we envision that biofunctional coatings might be relevant, especially in those medical applications where the implant is used for the treatment or therapy of a chronic pathology. Envisioned fields of interest also include neurodegenerative diseases and prosthetics for the central and peripheral nervous system. In particular, high-surface area metallic electrodes for neural recording and stimulation [106] may be realized by PED, even on highly flexible elastomeric substrates, and conductive and dielectric thin films for batteries, pacemakers, and devices based on organic bioelectronics interfacing with the human body [107].

Nevertheless, some crucial aspects concerning the quality of the coatings fabricated by PED still deserve deeper investigation. For instance, whereas crystalline coatings can be easily deposited by PED from target materials exhibiting simple chemical composition (i.e., ZrO_2) [15], this results in more challenges when complex or multiphase materials (i.e., calcium phosphates) are involved [58,73]. In these cases, the achievement of correct crystallinity (close to that of natural bone apatite), and therefore, also of optimal mechanical properties (coating toughness, hardness, and adhesion) still requires a post deposition thermal treatment. Obviously, this aspect is particularly relevant when the deposition is carried out at room temperature to allow the coating of soft and heat-sensitive materials [65].

In addition, some technological issues have still to be solved; for instance, the necessity of periodic cleaning of the auxiliary electrode in the IJD configuration or the relative costs for the installation of the equipment, albeit this feature is shared by all the main PVD techniques working in high vacuum. Finally, it should be remarked that PED is a line-of-sight technique, so the coating of complex shapes or 3D substrates is not trivial; in these cases, multi-step depositions or multi-axis rotating substrate holders must be considered to obtain homogeneous coatings.

7. Conclusions

In this work, we reviewed the recent technological developments and applications of PED in the biomedical field in light of its high versatility, mainly due to:

- The possibility to vary the depositions of a wide range of materials in a broad spectrum of operating independent conditions;
- The possibility to efficiently operate at low temperatures thanks to the high plasma density;
- The high fidelity in stoichiometry transfer from the target to the coating;
- The technological maturity and the competitive upscaling costs compared to similar techniques, such as PLD.

PED technology promises to conquer a remarkable role in the field of coating technology, even in the biomedical field where it can still be considered at its infancy.

Author Contributions: Conceptualization, M.B.; methodology, A.L., M.B. and C.G.; writing—original draft preparation, A.L. and M.B.; writing—review and editing, F.B., C.G. and M.L.F.; supervision, M.B. and C.G. All authors have read and agreed to the published version of the manuscript.

Funding: This research received no external funding.

Acknowledgments: The authors dedicate this work to the memory of Carlo Taliani, that pioneered PED for spintronics and coatings.

Conflicts of Interest: The authors declare no conflict of interest.

References

1. Wan, R.; Yang, M.; Zhou, Q.; Zhang, Q. Transparent conductive indium zinc oxide films prepared by pulsed plasma deposition. *J. Vac. Sci. Technol. A* **2012**, *30*, 61508. [CrossRef]
2. Arisi, E.; Bergenti, I.; Cavallini, M.; Murgia, M.; Riminucci, A.; Ruani, G.; Dediu, V. Room temperature deposition of magnetite thin films on organic substrate. *J. Magn. Magn. Mater.* **2007**, *316*, 410–412. [CrossRef]
3. Huang, Y.; Zhang, Q.; Li, G. Transparent conductive tungsten-doped tin oxide polycrystalline films prepared on quartz substrates. *Semicond. Sci. Technol.* **2009**, *24*, 15003. [CrossRef]
4. Huang, Y.W.; Li, G.F.; Feng, J.H.; Zhang, Q. Investigation on structural, electrical and optical properties of tungsten doped tin oxide thin films. *Thin Solid Films* **2010**, *518*, 1892–1896. [CrossRef]
5. Yao, Q.J.; Li, S.X.; Zhang, Q. Influences of channel metallic composition on indium zinc oxide thin-film transistor performance. *Semicond. Sci. Technol.* **2011**, *26*, 85011. [CrossRef]
6. Huang, Y.W.; Zhang, Q.; Xi, J.H.; Ji, Z.G. Transparent conductive p-type lithium-doped nickel oxide thin films deposited by pulsed plasma deposition. *Appl. Surf. Sci.* **2012**, *258*, 7435–7439. [CrossRef]
7. Yang, M.; Pu, H.F.; Zhou, Q.F.; Zhang, Q. Transparent p-type conducting K-doped NiO films deposited by pulsed plasma deposition. *Thin Solid Films* **2012**, *520*, 5884–5888. [CrossRef]
8. Witke, T.; Lenk, A.; Siemroth, P. Channel spark discharges for thin film technology. In Proceedings of the 17th International Symposium on Discharges and Electrical Insulation in Vacuum, Berkeley, CA, USA, 21–26 July 1996.
9. Witke, T.; Ziegele, H. Plasma state in pulsed arc, laser and electron deposition. *Surf. Coat. Technol.* **1997**, *97*, 414–419. [CrossRef]
10. Dediu, V.I.; Jiang, J.D.; Matacotta, F.C.; Scardi, P.; Lazzarino, M.; Nieva, G.; Civale, L. Deposition of $MBa_2Cu_3O_{7-x}$ thin films by channel-spark method. *Supercond. Sci. Technol.* **1995**, *8*, 160–164. [CrossRef]
11. Henda, R.; Wilson, G.; Gray-Munro, J.; Alshekhli, O.; McDonald, A.M. Preparation of polytetrafluoroethylene by pulsed electron ablation: Deposition and wettability aspects. *Thin Solid Films* **2012**, *520*, 1885–1889. [CrossRef]
12. Taliani, C.; Nozar, P. Method for Depositing Metal Oxide Films. U.S. Patent No. 12/809,326, 3 March 2011.
13. Lotti, R.; Nozar, P.; Taliani, C. Device for Generating Plasma and for Directing a Flow of Electrons towards a Target. U.S. Patent No. 8803425B2, 12 August 2014.

14. Bianchi, M.; Russo, A.; Taliani, C.; Marcacci, M. *Pulsed Plasma Deposition (PPD) Technique, in Dekker Encyclopedia of Nanoscience and Nanotechnology*, 3rd ed.; Taylor and Francis: New York, NY, USA, 2015; pp. 1–7.
15. Bianchi, M.; Russo, A.; Lopomo, N.; Boi, M.; Maltarello, M.C.; Sprio, S.; Baracchi, M.; Marcacci, M. Pulsed plasma deposition of zirconia thin films on UHMWPE: Proof of concept of a novel approach for joint prosthetic implants. *J. Mater. Chem. B* **2013**, *1*, 310–318. [CrossRef]
16. Harshavardhan, K.S.; Strikovski, M. *Second-Generation HTS Conductors*; Goyal, A., Ed.; Springer: New York, NY, USA, 2005; pp. 109–133.
17. Henda, R.; Al-Shareeda, O.; McDonald, A.; Pratt, A. Deposition of iron pyrite via pulsed electron ablation. *Appl. Phys. A* **2012**, *108*, 967–974. [CrossRef]
18. Strikovski, M.; Harshavardhan, K.S. Parameters that control pulsed electron beam ablation of materials and film deposition processes. *Appl. Phys. Lett.* **2003**, *82*, 853–855. [CrossRef]
19. Chandra, V.; Manoharan, S.S. Pulsed electron beam deposition of highly oriented thin films of polytetrafluoroethylene. *Appl. Surf. Sci.* **2008**, *254*, 4063–4066. [CrossRef]
20. Mathis, J.E.; Christen, H.M. Factors that influence particle formation during pulsed electron deposition of YBCO precursors. *Physica C* **2007**, *459*, 47–51. [CrossRef]
21. Christen, H.M.; Lee, D.F.; List, F.A.; Cook, S.W.; Leonard, K.J.; Heatherly, L.; Martin, P.M.; Paranthaman, P.; Goyal, A.; Rouleau, C.M. Pulsed electron deposition of fluorine-based precursors for $YBa_2Cu_3O_{7-x}$-coated conductors. *Supercond. Sci. Technol.* **2005**, *18*, 1168–1175. [CrossRef]
22. Skocdopolova, L. Device for Generating Plasma and Directing an Electron Beam towards a Target. WO Patent 2013/186697, 25 December 2013.
23. Mazzer, M.; Rampino, S.; Gombia, E.; Bronzoni, M.; Bissoli, F.; Pattini, F.; Calicchio, M.; Kingma, A.; Annoni, F.; Calestani, D.; et al. Progress on low-temperature pulsed electron deposition of $CuInGaSe_2$ solar cells. *Energies* **2016**, *9*, 207–218. [CrossRef]
24. Menossi, D.; Di Mare, S.; Rimmaudo, I.; Artegiani, E.; Tedeschi, G.; Pena, J.L.; Piccinell, F.; Salavei, F.; Romeo, A. SnS by Ionized Jet Deposition for photovoltaic applications. In Proceedings of the IEEE 44th Photovoltaic Specialist Conference (PVSC), Washington, DC, USA, 25–30 June 2017.
25. Skocdopole, J.; Aversa, L.; Golan, M.; Schenk, A.; Baldi, G.; Kratochvilova, I.; Kalvoda, L.; Nozar, P. Preparing of the chameleon coating by the ion jet deposition method. *Acta Polytech. CTU Proc.* **2017**, *9*, 19–25. [CrossRef]
26. Skocdopole, J.; Kalvoda, L.; Nozar, P.; Netopilik, M. Preparation of polymeric coatings by ionized jet deposition method. *Chem. Pap.* **2018**, *72*, 1735–1739. [CrossRef]
27. Zenker, R. *Electron Beam Surface Technologies*; Wang, Q.J., Chung, Y.W., Eds.; Springer: Boston, MA, USA, 2013.
28. Duta, L.; Popescu, A.C. Current status on pulsed laser deposition of coatings from animal-origin calcium phosphate sources. *Coatings* **2019**, *9*, 335–376. [CrossRef]
29. Shia, J.Z.; Chena, C.Z.; Yub, H.J.; Zhang, S.J. The effect of process conditions on the properties of bioactive films prepared by magnetron sputtering. *Vacuum* **2008**, *83*, 249–256. [CrossRef]
30. Pelletier, J.; Anders, A. Plasma-based ion implantation and deposition: A review of physics, technology, and applications. *IEEE Trans. Plasma Sci.* **2005**, *33*, 1944–1959. [CrossRef]
31. Peng, C.; Zhao, Y.; Jin, S.; Wang, J.R.; Liu, R.; Liu, H.; Shi, W.; Kolawole, K.S.; Ren, L.; Yu, B.; et al. Antibacterial TiCu/TiCuN multilayer films with good corrosion resistance deposited by axial magnetic field-enhanced arc ion plating. *ACS Appl. Mater. Interfaces* **2019**, *11*, 125–136. [CrossRef] [PubMed]
32. Döring, J.; Crackau, M.; Nestler, C.; Welzel, F.; Bertrand, J.; Lohmann, C.H. Characteristics of different cathodic arc deposition coatings on CoCrMo for biomedical applications. *J. Mech. Behav. Biomed. Mater.* **2019**, *97*, 212–221. [CrossRef] [PubMed]
33. Sobieraj, M.C.; Rimnac, C.M. Ultra high molecular weight polyethylene: Mechanics, morphology, and clinical behavior. *J. Mech. Behav. Biomed. Mater.* **2009**, *2*, 433–443. [CrossRef] [PubMed]
34. Dumbleton, J.H.; Manley, M.T.; Edidin, A.A. A literature review of the association between wear rate and osteolysis in total hip arthroplasty. *J. Arthroplast.* **2002**, *17*, 649–661. [CrossRef]
35. Ingham, E.; Fisher, J. The role of macrophages in osteolysis of total joint replacement. *Biomaterials* **2005**, *26*, 1271–1286. [CrossRef]
36. Drees, P.; Eckardt, A.; Gay, R.E.; Gay, S.; Huber, L.C. Mechanisms of disease: Molecular insights into aseptic loosening of orthopedic implants. *Nat. Clin. Pract. Rheumatol.* **2007**, *3*, 165–171. [CrossRef]

37. Kaufman, A.M.; Alabre, C.I.; Rubash, H.E.; Shanbhag, A.S. Human macrophage response to UHMWPE, TiAlV, CoCr, and alumina particles: analysis of multiple cytokines using protein arrays. *J. Biomed. Mater. Res. Part A* **2008**, *84*, 464–474. [CrossRef]
38. Labek, G.; Thaler, M.; Janda, W.; Agreiter, M.; Stöckl, B. Revision rates after total joint replacement: Cumulative, Cumulative results from worldwide joint register datasets. *J. Bone Joint Surg. Br.* **2011**, *93*, 293–297. [CrossRef]
39. Russo, A.; Bianchi, M.; Lopomo, N.; Boi, M.; Ortolani, A.; Marchiori, G.; Gambardella, A.; Maltarello, M.C.; Visani, A.; Marcacci, M. Hard ceramic thin films on UHMWPE insert: A novel approach for longlasting joint implants. *Orthop. Proc.* **2016**, *98*, 10.
40. Bianchi, M.; Boi, M.; Lopomo, N. Nanomechanical characterization of zirconia thin films deposited on UHMWPE by pulsed plasma deposition. *J. Mech. Med. Biol.* **2015**, *15*, 1550070–1550085. [CrossRef]
41. Marchiori, G.; Lopomo, N.; Boi, M.; Berni, M.; Bianchi, M.; Gambardella, A.; Visani, A.; Russo, A.; Marcacci, M. Optimizing thickness of ceramic coatings on plastic components for orthopedic applications: A finite element analysis. *Mater. Sci. Eng. C* **2016**, *58*, 381–388. [CrossRef] [PubMed]
42. Berni, M.; Marchiori, G.; Gambardella, A.; Boi, M.; Bianchi, M.; Russo, A.; Visani, A.; Marcacci, M.; Pavanc, P.G.; Lopomo, N.F. Effects of working gas pressure on zirconium dioxide thin film prepared by pulsed plasma deposition: Roughness, wettability, friction and wear characteristics. *J. Mech. Behav. Biomed. Mater* **2017**, *72*, 200–208. [CrossRef] [PubMed]
43. Berni, M.; Lopomo, N.; Marchiori, G.; Gambardella, A.; Boi, M.; Bianchi, M.; Visani, A.; Pavan, P.; Russo, A.; Marcacci, M. Tribological characterization of zirconia coatings deposited on Ti6Al4V components for orthopedic applications. *Mater. Sci. Eng.* **2016**, *62*, 643–655. [CrossRef] [PubMed]
44. Bianchi, M.; Lopomo, N.; Boi, M.; Gambardella, A.; Marchiori, G.; Berni, M.; Pavan, P.; Marcacci, M.; Russo, A. Ceramic thin films realized by means of pulsed plasma deposition technique: Applications for orthopedics. *J. Mech. Med. Biol.* **2015**, *15*, 1540002–1540009. [CrossRef]
45. Bianchi, M.; Gambardella, A.; Berni, M.; Panseri, S.; Montesi, M.; Lopomo, N.; Tampieri, A.; Marcacci, M.; Russo, A. Surface morphology, tribological properties and in vitro biocompatibility of nanostructured zirconia thin films. *J. Mater. Sci. Mater. Med.* **2016**, *27*, 95–105. [CrossRef]
46. Dong, H.; Shi, W.; Bell, T. Potential of improving tribological performance of UHMWPE by engineering the Ti6Al4V counterfaces. *Wear* **1999**, *225*, 146–153. [CrossRef]
47. Xiong, D.; Ge, S. Friction and wear properties of UHMWPE/Al_2O_3 ceramic under different lubricating conditions. *Wear* **2001**, *250*, 242–245. [CrossRef]
48. Gambardella, A.; Berni, M.; Russo, A.; Bianchi, M. A comparative study of the growth dynamics of zirconia thin films deposited by ionized jet deposition onto different substrates. *Surf. Coat. Technol.* **2018**, *337*, 306–312. [CrossRef]
49. Kurtz, S.M.; Kocagoz, S.; Arnholt, C.; Huet, R.; Ueno, M.; Walter, W.L. Advances in zirconia toughened alumina biomaterials for total joint replacement. *J. Mech. Behav. Biomed. Mater.* **2014**, *31*, 107–116. [CrossRef] [PubMed]
50. De Aza, A.H.; Chevalier, J.; Fantozzi, G.; Schehl, M.; Torrecillas, R. Crack growth resistance of alumina, zirconia and zirconia toughened alumina ceramics for joint prostheses. *Biomaterials* **2002**, *23*, 937–945. [CrossRef]
51. Russo, A.; Lopomo, N.; Bianchi, M.; Boi, M.; Ortolani, A.; Gambardella, A.; Marchiori, G.; Maltarello, M.C.; Visani, A.; Marcacci, M. Protective zirconia-toughened-alumina films deposited on the UHMWPE insert by the pulsed plasma deposition method: Preliminary results. *Orthop. Proc.* **2018**, *98*, 10.
52. Kaciulis, S.; Mezzi, A.; Bianchi, M.; Gambardella, A.; Boi, M.; Liscio, F.; Marcacci, M.; Russo, A. Ceramic coatings for orthopaedic implants: Preparation and characterization. *Surf. Interface Anal.* **2015**, *48*, 616–620. [CrossRef]
53. Junker, R.; Dimakis, A.; Thoneick, M.; Jansen, J.A. Effects of implant surface coatings and composition on bone integration: A systematic review. *Clin. Oral Implants Res.* **2009**, *20*, 185–206. [CrossRef]
54. Liu, X.; Chu, P.K.; Ding, C. Surface modification of titanium, titanium alloys, and related materials for biomedical applications. *Mater. Sci. Eng. R Rep.* **2004**, *57*, 49–121. [CrossRef]
55. Nijhuis, A.W.; Leeuwenburgh, S.C.; Jansen, J.A. Wet-chemical deposition of functional coatings for bone implantology. *Macromol. Biosci.* **2010**, *10*, 1316–1329. [CrossRef]

56. Surmenev, R.A. A review of plasma-assisted methods for calcium phosphate-based coatings fabrication. *Surf. Coat. Technol.* **2012**, *206*, 2035–2056. [CrossRef]
57. Surmenev, R.A.; Surmeneva, M.A.; Ivanova, A.A. Significance of calcium phosphate coatings for the enhancement of new bone osteogenesis—A review. *Acta Biomater.* **2014**, *10*, 557–579. [CrossRef]
58. Boi, M.; Bianchi, M.; Gambardella, A.; Liscio, F.; Kaciulis, S.; Visani, A.; Barbalinardo, M.; Valle, F.; Iafisco, M.; Lungaro, L.; et al. Tough and adhesive nanostructured calcium phosphate thin films deposited by the pulsed plasma deposition method. *RSC Adv.* **2015**, *5*, 78561–78571. [CrossRef]
59. Kurtz, S.M.; Devine, J.N. PEEK biomaterials in trauma, orthopedic, and spinal implants. *Biomaterials* **2007**, *28*, 4845–4869. [CrossRef] [PubMed]
60. Du, Y.W.; Zhang, L.N.; Hou, Z.T.; Ye, X.; Gu, H.S.; Yan, G.P.; Shang, P. Physical modification of polyetheretherketone for orthopedic implants. *Front. Mater. Sci.* **2014**, *8*, 313–324. [CrossRef]
61. Toth, J.M.; Wang, M.; Estes, B.T.; Scifert, J.L.; Seim Iii, H.B.; Turner, A.S. Polyetheretherketone as a biomaterial for spinal applications. *Biomaterials* **2006**, *27*, 324–334. [CrossRef] [PubMed]
62. Williams, D.F.; McNamara, A.; Turner, R.M. Potential of polyetheretherketone (PEEK) and carbon-fibre-reinforced PEEK in medical applications. *J. Mater. Sci. Lett.* **1987**, *6*, 188–190. [CrossRef]
63. Zhao, M.; An, M.; Wang, Q.; Liu, X.; Lai, W.; Zhao, X.; Wei, S.; Ji, J. Quantitative proteomic analysis of human osteoblast-likeMG-63 cells in response to bioinert implant material titanium and polyetheretherketone. *J. Proteome* **2012**, *75*, 3560–3573. [CrossRef]
64. Sagomonyants, K.B.; Jarman-Smith, M.L.; Devine, J.N.; Aronow, M.S.; Gronowicz, G.A. The in vitro response of human osteoblasts to polyetheretherketone (PEEK) substrates compared to commercially pure titanium. *Biomaterials* **2008**, *29*, 1563–1572. [CrossRef]
65. Bianchi, M.; Degli Esposti, L.; Ballardini, A.; Liscio, F.; Berni, M.; Gambardella, A.; Leeuwenburghd, S.G.C.; Sprio, S.; Tampieri, A.; Iafisco, M. Strontium doped calcium phosphate coatings on poly(etheretherketone) (PEEK) by pulsed electron deposition. *Surf. Coat. Technol.* **2017**, *319*, 191–199. [CrossRef]
66. Marie, P.J. Strontium as therapy for osteoporosis. *Curr. Opin. Pharmacol.* **2005**, *5*, 633–636. [CrossRef]
67. Braceras, I.; Vera, C.; Izquierdo, A.A.; Muñoz, R.; Lorenzo, J.; Alvareza, N.; Maeztu, M.A. Ion implantation induced nanotopography on titanium and bone cell adhesion. *Appl. Surf. Sci.* **2014**, *310*, 24–30. [CrossRef]
68. Deligianni, D.D.; Katsala, N.D.; Koutsoukos, P.G.; Missirlis, Y.F. Effect of surface roughness of hydroxyapatite on human bone marrow cell adhesion, proliferation, differentiation and detachment strength. *Biomaterials* **2000**, *22*, 87–96. [CrossRef]
69. Boanini, E.; Gazzano, M.; Bigi, A. Ionic substitutions in calcium phosphates synthesized at low temperature. *Acta Biomater.* **2010**, *6*, 1882–1894. [CrossRef] [PubMed]
70. Graziani, G.; Bianchi, M.; Sassoni, E.; Russo, A.; Marcacci, M. Ion-substituted calcium phosphate coatings deposited by plasma-assisted techniques: A review. *Mater. Sci. Eng. C* **2017**, *74*, 219–229. [CrossRef] [PubMed]
71. Graziani, G.; Boi, M.; Bianchi, M. A review on ionic substitutions in hydroxyapatite thin films: Towards complete biomimetism. *Coatings* **2018**, *8*, 269–285. [CrossRef]
72. Rau, J.V.; Cacciotti, I.; Laureti, S.; Fosca, M.; Varvaro, G.; Latini, A. Bioactive, nanostructured Si-substituted hydroxyapatite coatings on titanium prepared by pulsed laser deposition. *J. Biomed. Mater. Res. B Appl. Biomater.* **2015**, *103*, 1621–1631. [CrossRef] [PubMed]
73. Bianchi, M.; Gambardella, A.; Graziani, G.; Liscio, F.; Maltarello, M.C.; Boi, M.; Berni, M.; Bellucci, D.; Marchiori, G.; Valle, F.; et al. Plasma-assisted deposition of bone apatite-like thin films from natural apatite. *Mater. Lett.* **2017**, *199*, 32–36. [CrossRef]
74. Graziani, G.; Berni, M.; Gambardella, A.; De Carolis, M.; Cristina Maltarello, M.; Boi, M.; Carnevale, G.; Bianchi, M. Fabrication and characterization of biomimetic hydroxyapatite thin films for bone implants by direct ablation of a biogenic source. *Mater. Sci. Eng. C* **2019**, *99*, 853–862. [CrossRef]
75. Miller, L.M.; Vairavamurthy, V.; Chance, M.R.; Mendelsohn, R.; Paschalis, E.P.; Betts, F.; Boskey, A.L. In situ analysis of mineral content and crystallinity in bone using infrared micro-spectroscopy of the ν_4 PO_4^{3-} vibration. *Biochim. Biophys. Acta* **2001**, *1527*, 11–19. [CrossRef]
76. Bianchi, M.; Pisciotta, A.; Bertoni, L.; Berni, M.; Gambardella, A.; Visani, A.; Russo, A.; de Pol, A.; Carnevale, G. Osteogenic differentiation of hDPSCs on biogenic bone apatite thin films. *Stem Cells Int.* **2017**, *2017*, 1–10.
77. Hench, L.L. The story of Bioglass®. *J. Mater. Sci. Mater. Med.* **2006**, *17*, 967–978. [CrossRef]

78. Bellucci, D.; Bianchi, M.; Graziani, G.; Gambardella, A.; Berni, M.; Russo, A.; Cannillo, V. Pulsed electron deposition of nanostructured bioactive glass coatings for biomedical applications. *Ceram. Int.* **2017**, *43*, 15862–15867. [CrossRef]
79. Vafa, H.M.; Johansson, D.; Wallen, H.; Sanchez, J. Improved ex vivo blood compatibility of central venous catheter with noble metal alloy coating. *J. Biomed. Mater. Res. B Appl. Biomater.* **2016**, *104*, 1359–1365. [CrossRef] [PubMed]
80. Veerachamy, S.; Yarlagadda, T.; Manivasagam, G.; Yarlagadda, P.K. Bacterial adherence and biofilm formation on medical implants: A review. *Proc. Inst. Mech. Eng. H* **2014**, *228*, 1083–1099. [CrossRef] [PubMed]
81. Zilberman, M.; Elsner, J.J. Antibiotic-eluting medical devices for various applications. *J. Control. Release* **2008**, *130*, 202–215. [CrossRef] [PubMed]
82. Alt, V. Antimicrobial coated implants in trauma and orthopaedics-A clinical review and risk-benefit analysis. *Injury* **2017**, *48*, 599–607. [CrossRef]
83. Blaker, J.J.; Nazhat, S.N.; Boccaccini, A.R. Development and characterisation of silver-doped bioactive glass-coated sutures for tissue engineering and wound healing applications. *Biomaterials* **2004**, *25*, 1319–1329. [CrossRef]
84. Branemark, P. Tissue-Integrated Prostheses. Osseointegration in Clinical Dentistry. In *Introduction to Osseointegration*; Branemark, P.I., Zarb, C., Albrektsson, T., Eds.; Quintessence Publishing Co.: Chicago, IL, USA, 1985; pp. 11–76.
85. Raphel, J.; Holodniy, M.; Goodman, S.B.; Heilshorn, S.C. Multifunctional coatings to simultaneously promote osseointegration and prevent infection of orthopaedic implants. *Biomaterials* **2016**, *84*, 301–314. [CrossRef]
86. Xie, C.M.; Lu, X.; Wang, K.F.; Meng, F.Z.; Jiang, O.; Zhang, H.P.; Zhi, W.; Fang, L.M. Silver nanoparticles and growth factors incorporated hydroxyapatite coatings on metallic implant surfaces for enhancement of osteoinductivity and antibacterial properties. *ACS Appl. Mater. Interfaces* **2014**, *6*, 8580–8589. [CrossRef]
87. Kolmas, J.; Groszyk, E.; Kwiatkowska-Różycka, D. Substituted hydroxyapatites with antibacterial properties. *Biomed. Res. Int.* **2014**, *2014*, 1–15. [CrossRef]
88. Melaiye, A.; Youngs, W.J. Silver and its application as an antimicrobial agent. *Expert Opin. Ther. Pat.* **2005**, *15*, 125–130. [CrossRef]
89. Galdiero, S.; Falanga, A.; Vitiello, M.; Cantisani, M.; Marra, M.; Galdiero, M. Silver nanoparticles as potential antiviral agents. *Molecules* **2011**, *16*, 8894–8918. [CrossRef]
90. Lansdown, A.B.G. A pharmacological and toxicological profile of Ag as an antimicrobial agent in medical devices. *Adv. Phar. Sci.* **2010**, *2010*, 1–16. [CrossRef] [PubMed]
91. Pauksch, L.; Hartmann, S.; Rohnk, M.; Szalay, G.; Alt, V.; Schnettler, R.; Lips, K.S. Biocompatibility of Ag nanoparticles and Ag ions in primary human mesenchymal stem cells and osteoblasts. *Acta Biomater.* **2014**, *10*, 439–449. [CrossRef] [PubMed]
92. Le Ouay, B.; Stellacci, F. Antibacterial activity of Ag nanoparticles: A surface science insight. *Nano Today* **2015**, *10*, 339–354. [CrossRef]
93. Gambardella, A.; Berni, M.; Graziani, G.; Kovtun, A.; Liscio, A.; Russo, A.; Visani, A.; Bianchi, M. Nanostructured Ag thin films deposited by pulsed electron ablation. *Appl. Surf. Sci.* **2019**, *475*, 917–925. [CrossRef]
94. Li, B.; Webster, T. *Orthopedic Biomaterials: Advances and Applications*; Springer: Berlin, Germany, 2018.
95. Ficai, A.; Grumezescu, A. *Nanostructures for Antimicrobial Therapy*, 1st ed.; Elsevier: Amsterdam, The Netherlands, 2017.
96. Thompson, C.V. Solid-state dewetting of thin films. *Annu. Rev. Mater. Res.* **2012**, *42*, 399–434. [CrossRef]
97. Jacquet, P.; Podor, R.; Ravaux, J.; Teisseire, J.; Gozhyk, I.; Jupille, J.; Lazzari, R. Grain growth: The key to understand solid-state dewetting of silver thin films. *Scripta Mater.* **2016**, *115*, 128–132. [CrossRef]
98. Leroy, F.; Borowik, L.; Cheynis, F.; Almadori, Y.; Curiotto, S.; Trautmann, M.; Barbè, J.C.; Muller, P. How to control solid state dewetting: A short review. *Surf. Sci. Rep.* **2016**, *71*, 391–409. [CrossRef]
99. Berni, M.; Carrano, I.; Kovtun, A.; Russo, A.; Visani, A.; Dionigi, C.; Liscio, A.; Valle, F.; Gambardella, F. Monitoring morphological and chemical properties during silver solid-state dewetting. *Appl. Surf. Sci.* **2019**, *498*, 1–8. [CrossRef]
100. Taylor, E.N.; Webster, T.J. The use of superparamagnetic nanoparticles for prosthetic biofilm prevention. *Int. J. Nanomed.* **2009**, *4*, 145–152.

101. Gambardella, A.; Bianchi, M.; Kaciulis, S.; Mezzi, A.; Brucale, M.; Cavallini, M.; Herrmannsdoerfer, T.; Chanda, G.; Uhlarz, M.; Cellini, A.; et al. Magnetic hydroxyapatite coatings as a new tool in medicine: A scanning probe investigation. *Mater. Sci. Eng. C* **2016**, *62*, 444–449. [CrossRef]
102. Amberg, M.; Grieder, K.; Barbadoro, P.; Heuberger, M.; Hegemann, D. Electromechanical behavior of nanoscale silver coatings on PET fibers. *Plasma Process. Polym.* **2008**, *5*, 874–880. [CrossRef]
103. Atwa, Y.; Maheshwari, N.; Goldthorpe, I.A. Silver nanowire coated threads for electrically conductive textiles. *J. Mater. Chem. C* **2015**, *3*, 3908–3912. [CrossRef]
104. Ortolani, A.; Bianchi, M.; Mosca, M.; Caravelli, S.; Fuiano, M.; Marcacci, M.; Russo, A. The prospective opportunities offered by magnetic scaffolds for bone tissue engineering: A review. *Joints* **2016**, *4*, 228–235. [CrossRef] [PubMed]
105. Vandrovcova, M.; Douglasb, T.E.L.; Mróz, W.; Musial, O.; Schaubroeck, D.; Budner, B.; Syroka, R.; Dubruel, P.; Bacakova, L. Pulsed laser deposition of magnesium-doped calcium phosphate coatings on porous polycaprolactone scaffolds produced by rapid prototyping. *Mater. Lett.* **2015**, *148*, 178–183. [CrossRef]
106. Carli, S.; Bianchi, M.; Zucchini, E.; Di Lauro, M.; Prato, M.; Murgia, M.; Fadiga, L.; Biscarini, F. Electrodeposited PEDOT: Nafion composite for neural recording and stimulation. *Adv. Healthc. Mater.* **2019**, *8*, 1900765–1900775. [CrossRef]
107. Di Lauro, M.; Benaglia, S.; Berto, M.; Bortolotti, C.A.; Zoli, M.; Biscarini, F. Exploiting interfacial phenomena in organic bioelectronics: Conformable devices for bidirectional communication with living systems. *Colloids Surf. B Biointerfaces* **2018**, *168*, 143–147. [CrossRef]

MDPI
St. Alban-Anlage 66
4052 Basel
Switzerland
Tel. +41 61 683 77 34
Fax +41 61 302 89 18
www.mdpi.com

Coatings Editorial Office
E-mail: coatings@mdpi.com
www.mdpi.com/journal/coatings

www.ingramcontent.com/pod-product-compliance
Lightning Source LLC
LaVergne TN
LVHW070900170726
843515LV00005B/689
* 9 7 8 3 0 3 6 5 2 4 1 4 6 *